CELL SUBSTRATES

Their Use in the Production of Vaccines and Other Biologicals

Edited by

John C. Petricciani
and Hope E. Hopps

Bureau of Biologics of the Food and Drug Administration
Bethesda, Maryland

and

Paul J. Chapple

W. Alton Jones Cell Science Center
Lake Placid, New York

PLENUM PRESS • NEW YORK AND LONDON

Library of Congress Cataloging in Publication Data

Symposium on Cell Substrates and Their Use in the Production of Vaccines and Other Biologicals, 2d, Lake Placid, N. Y., 1978.
Cell substrates, their use in the production of vaccines and other biologicals.

(Advances in experimental medicine and biology; v. 118).
"Proceedings of the Second Annual Symposium on Cell Substrates and Their Use in the Production of Vaccines and Other Biologicals, held October 23-26, 1978."
"Symposium sponsored by the W. Alton Jones Cell Science Center."
Includes index.
1. Vaccines – Congresses. 2. Human cell culture – Congresses. 3. Biological products – Congresses. 4. Cancer cells – Congresses. I. Petricciani, John C. II. Hopps, Hope E. III. Chapple, Paul J. IV. Tissue Culture Association. W. Alton Jones Cell Science Center. V. Title. VI. Series.
QR189.S93 1978 574'.07'24 79-13479
ISBN 978-1-4684-0999-4 ISBN 978-1-4684-0997-0 (eBook)
DOI 10.1007/978-1-4684-0997-0

Proceedings of the Second Annual Symposium, on Cell Substrates and Their Use in the Production of Vaccines and Other Biologicals, held at the W. Alton Jones Cell Science Center, Lake Placid, New York, October 23–26, 1978.

Symposium sponsored by the Tissue Culture Association, Inc., and its operational unit, the W. Alton Jones Cell Science Center, with the support of:

Bureau of Biologics, FDA
Fogarty International Center, National Institutes of Health
National Institute of Allergy and
Infectious Diseases
SYNTEX

Softcover reprint of tht heardcover 1st edition 1979
A Division of Plenum Publishing Corporation
227 West 17th Street, New York, N.Y. 10011

CELL SUBSTRATES
Their Use in the Production of Vaccines and Other Biologicals

ADVANCES IN EXPERIMENTAL MEDICINE AND BIOLOGY

Recent Volumes in this Series

Volume 109
DRUGS, LIPID METABOLISM, AND ATHEROSCLEROSIS
Edited by David Kritchevsky, Rodolfo Paoletti, and William L. Holmes

Volume 110
HUMAN INTERFERON: Production and Clinical Use
Edited by Warren R. Stinebring and Paul J. Chapple

Volume 111
HORMONES AND ENERGY METABOLISM
Edited by David M. Klachko, Ralph R. Anderson, and Murray Heimberg

Volume 112
OVARIAN FOLLICULAR AND CORPUS LUTEUM FUNCTION
Edited by Cornelia P. Channing, John M. Marsh, and William A. Sadler

Volume 113
PARKINSON'S DISEASE –II: Aging and Neuroendocrine Relationships
Edited by Caleb E. Finch, David E. Potter, and Alexander D. Kenny

Volume 114
FUNCTION AND STRUCTURE OF THE IMMUNE SYSTEM
Edited by Wolfgang Müller-Ruchholtz and Hans Konrad Müller-Hermelink

Volume 115
DYNAMICS OF ARTERIAL FLOW
Edited by Stewart Wolf and Nicholas T. Werthessen

Volume 116
MODULATORS, MEDIATORS, AND SPECIFIERS IN BRAIN FUNCTION
Edited by Yigal H. Ehrlich, Jan Volavka, Leonard G. Davis, and Eric G. Brunngraber

Volume 117
STEROID HORMONE RECEPTOR SYSTEMS
Edited by Wendell W. Leavitt and James H. Clark

Volume 118
CELL SUBSTRATES: Their Use in the Production of Vaccines and Other Biologicals
Edited by John C. Petricciani, Hope E. Hopps, and Paul J. Chapple

Volume 119
TREATMENT OF EARLY DIABETES
Edited by Rafael A. Camerini-Davalos and Bernard Hanover

Preface

This volume stems from a symposium sponsored by the W. Alton Jones Cell Science Center, Lake Placid New York. The Second Annual W. Alton Jones Cell Science Center Symposium: Cell Substrates and Their Use in the Production of Vaccines and Other Biologicals was held October 23-26, 1978. The Center is an operational unit of the Tissue Culture Association and offers, in collaboration with the Association's Education Committee, a wide range of educational and research activities.

During the past 20 years there have been numerous national and international conferences on the topic of cell cultures used to produce biological products. Those largely dealt with the technology and associated issues that were current at the time of the meetings. For example, as human diploid cells were developed and proposed for use in vaccine production, a number of meetings were held to examine the pros and cons of human diploid cells. A large amount of data was provided at those conferences which formed the basis for the eventual acceptance of that cell system. Each meeting added to the general base of knowledge in the area of cell cultures and their application to the current and novel set of problems encountered. In general, the participants reaffirmed the basic premises that were formulated in the early days of polio virus vaccine production regarding the criteria for acceptability of cells when used in the manufacture of biologics intended for humans.

The present symposium follows the tradition of its predecessors in that we have included presentations related to current technology and to new biological products which can be produced in cell culture systems. We were concerned not only with the practical aspects of cell substrates and production of biologicals but also with the philosophical and ethical considerations in the types of substrates used and the manner in which they are used. The use of plant cells for the production of drugs, flavors, enzymes and colorings was one major omission from the program. Because this is an area which is developing rapidly and its potential is immense, we asked a leading expert in this field, Dr. Donald K. Dougall, to contribute a paper to this volume.

A unique feature of this symposium is a re-examination of some of the traditional concepts that have formed the basis for cell culture use in the production of biologicals up to the present time. The emergence of new experimental products such as interferon produced in lymphoblastoid cells has led us to re-examine some of the old dogmas concerning cell acceptability. As in any area of science, such reassessments can only be viewed as positive elements in the growth and development of the discipline.

In conjunction with this symposium, a meeting of the ad hoc Karyology Committee was held to review and revise the current recommendations for cytogenetic monitoring of human cell cultures used to produce biological products. The meeting took place immediately after the Symposium and many of the points discussed during the preceeding days were considered. The Committtee's report is included in the proceedings because of the direct relationship of karyology to the topics of this symposium, and because many of the symposium participants also attended the Committee meetings and helped to formulate the new recommendations.

By providing a forum for the free and open discussion of such topics as the use of tumorigenic cells or continuous cell lines, we hope to begin the process by which logical and reasonable decisions can be made based on data and experience rather than on personal bias and emotion. This symposium marks the first step along the path towards reaching a scientific concensus on the use of various cells for the production of both old and new human biologicals. If it serves to act as the stimulus for the development of new data which leads to the wider availability of products needed to treat or prevent human disease, then it will have fulfilled its mission.

ACKNOWLEDGEMENTS

The editors wish to acknowledge the support and assistance of the Education Committee of the Tissue Culture Association and the financial support of the Bureau of Biologics of the Food and Drug Administration, the Forgarty International Center, National Institute of Allergy and Infectious Diseases, and SYNTEX Agri Business, Animal Health Division. This volume would not have been possible without the skilled assistance of Mrs. Ellen Anastos-Dorchak and Ms. Marion Thomas in collecting, collating and retyping the manuscripts.

Contents

PAST ACHIEVEMENTS - FUTURE EXPECTATIONS

Harry M. Meyer, Jr., M.D.

Bureau of Biologics
8800 Rockville Pike
Bethesda, MD 20014

I would like to join Dr. Chapple in welcoming you to this Second Annual W. Alton Jones Cell Science Center Symposium. Also, speaking for the participants, I would like to express our appreciation for the hard work of Dr. Chapple and his staff in making this conference possible.

Long ago, mankind found periodic withdrawal from the frenzy of day to day activities helpful in taking the longer view of problems and progress. There have been and will continue to be many technical meetings on topics relating to cell cultures. It is my hope, however, that in coming to Lake Placid, we have come to a retreat - from which we can take the larger view. If this proves to be the case we will find ourselves indebted to the late Dr. George Gey and to Mrs. W. Alton Jones, who through their wisdom and generosity created the Center where people of various disciplines might come together to talk, learn and work in collectively dealing with the science of cell cultures.

There is a time for all things and what better time could be found for an overview than after 25 years experience in the use of cell cultures for the production of biologics? Scientists seek absolute answers but the critical role of the scientist in contemporary society is that of providing informed, expert guidance in a technically complex world in which few absolute answers exist. Our modern conquest of particular diseases has been the repeated story of measured progress in the face of incomplete knowledge. Because of this truth there are a couple of general concerns that I would like to mention in setting the tone for this conference. The first has to do with the role of the scientist and the regulator in the decision-making process. To cite an example of the problem, we continue to hear the question - "Will the Bureau of Biologics approve this or that?" Perhaps, in an increasingly procedural world it becomes easy for you to assume that he who proposes in the Federal Register has much in common with he who carves commandments on tablets of stone, yet such is not the case. While there is a statutory base of authority, in day to day practice a regulatory organization can do no more than follow the consensus in the informed public. For practical purposes that means that you, the scientists, dictate the decisions. Our chief role is one of being an expediter in helping you focus on a problem. Our major difficulty in this role comes from the fact that decisions are rarely unanimous, so that we, not you, come under fire from vocal minorities, as we implement the collective best judgement of the majority. While it is true that the regulator could tilt the process, Darwin's laws strongly favor the survival of regulators, that as a species, stay tuned to majority opinion.

The second general concern I wish to mention before turning to the specific issue we face is that of the social price of regulation. After six years, three months and 23 days in harness, I am particularly concerned about the potential for over-regulation of science. Speaking as a scientist and as a regulator - I feel it is fair to say that the scientific community has been remarkably naive about the regulatory process. Because of this, I have watched with interest, but from a distance, the dialogue on regulation of DNA research. Perhaps the enduring benefit of these debates will prove to be the education of some of our country's best basic scientists in the practical realities of the social costs of regulation. Who would not have their cake and eat it too, if that were an option? But how can we expect public understanding, if scientists, themselves, fail to recognize that innovation, by its nature, involves risk - and that absolute safety comes at a price of no innovation? This dilemma and the new maturation of the science community was well expressed in Dr. James Watson's Annual Report of the Cold Spring Harbour Laboratory: "...I, however, left Asilomar feeling very uneasy. My view that the whole affair was a hasty rush into unjustified bureaucratic roadblocks that would set back the course of legitimate science was considered by some to be eccentric irresponsibility. Yet I did not then, nor do I now, believe all recombinant DNA research is necessarily totally safe. The future automatically entails uncertainty, and the days of all of us are marked by unexpected occurrences that might suddenly do us in. Obviously, no sane person rushes in directions where he anticipates harm to himself or others, and we would be deluding ourselves if we were to proceed as if in a risk-free world. Instead, our sanity demands that we try to find means to order actions in terms of the magnitude of their potential risk. In particular, when no quantitation is possible because we have never faced a particular situation before, we must not assume the worst. If we did, we would quickly be suffocated by the inaction of unlimited prudent behavior."

Let me turn now for a few moments to the specific subject of the conference. From the biologics point of view it all began about 25 years ago with the realization that the new cell culture technology might be used in conquering poliomyelitis. How appropriate that here today as a participant, we have Dr. Salk, the architect of that first "cell-associated" medical crusade. With primary monkey kidney cell cultures and polio virus came the multitude of problems relating to extraneous agents - agents in the cells themselves, agents in bovine serum and trypsin, and agents introduced during manipulation of the cell cultures. By the late 1950's, both measles and polioviruses had been attenuated, creating the potential for live virus vaccines produced in chick embryo and monkey kidney cell cultures, respectively. With the entering of the 1960's, two types of events occurred that were to exert considerable influence over the use of cell cultures for biologics production. One was the clear statement of a majority opinion developed informally during the decade of the 50's. This opinion held that continuous cell lines, particularly if derived from neoplastic tissues presented more unknown risks than primary cell cultures, and hence, were unsuitable for biologics production. The other event concerned recognition that many animal species harbored viruses capable of producing malignant tumors in animals and that these viruses often could transform human cells. Equally startling was recognition that these viruses or genetic material from these viruses were unsuspected residents in many commonly used cell cultures. The effect of the recognition of avian leukosis viruses in experimental measles vaccine and of simian virus -40 (SV_{40}) in licensed and experimental polio vaccines was profound.

Just as these events sowed the seeds for the multimillion dollar virus Cancer Program, in the vaccine development field it nurtured a feeling of uneasiness and reluctance for any actions that represented a departure from established precedent. Thus, chick embryo cells were virtually enshrined, since their use, either in cultures or in the shell, won any length of practice contest, "hands down." Monkey kidney cells were viewed less reverently not only because of fears related to SV_{40}, Marburg and other contaminants, but also because of the lesser human experience with monkey cells. Maurice Hilleman pointed out, for example, that "raw monkey kidney was not a normal part of the human diet." Awareness of their problems as the decade of the 1960's progressed served as an impetus to research on diploid cells and to the development of closed flocks or colonies of pathogen-free animals as donors of tissues for vaccine production.

Another surprise of the late 1960's extending into the current decade came from the recognition that contamination of cell cultures with transformed cells was an all too common event. Thus, as a result of this collective wisdom human diploid cells were approved for vaccine production in 1970, but only with ongoing extraordinary testing, and all nonhuman primary cells used for vaccines (except monkey cells) came from closed locks or colonies. Now as we approach the end of the 1970's it seems likely that even monkey cells will come from closed colonies within the forseeable future.

Other discoveries of the past 10 to 20 years served to broaden the role of cell technology in the production of biologics. After a slow, sputtering start, reminiscent of the early days of penicillin, interferon seems to be on the verge of finding a place in medicine. Urokinase and other enzymes and hormones are likewise extractable from cultured cells. These new products have raised new issues relating to the degree to which the product may be purified, the increasingly acute need for new cell sources which do not meet traditional criteria of acceptability.

What have we learned about risk in the past 25 years that could help us in facing the future? In a sense we have been both frightened and reassured. Early concerns about tissue sensitization have not materialized. Kidney cell vaccines have not caused kidney damage in recipients. Anaphylaxis and other allergic manifestations appear to be virtually unseen since penicillin was removed from cell culture vaccines 20 years ago. The past problems with inactivated measles and respiratory syncytial vaccines did not appear to be cell culture related. Extraneous agents have, on occasion, been administered inadvertently to many vaccine recipients but no evidence of harm has been forthcoming. In fact, in a sense, it is amazing to be able to say that after 25 years, and millions and millions of doses of cell culture derived vaccines we are unaware of any documented harm to any recipient that was a result of the cell culture itself or an extraneous agent contained in the culture.

In closing, I would like to ask that we look carefully at the experience of the past to determine what is and what is not reasonable, in terms of controls, as we move into the future. Have some of our rules and "no-no's" been dictated more by the fears and uncertainties of the times than by reason?

- What, for example, did we learn by requiring that 100,000 persons be inoculated with a vaccine prepared in dog kidney before we accepted dog kidney cell cultures?

- What did we get from the regulation calling for an experience with 10,000 or more persons inoculated with WI-38 cell-produced polio vaccine? We have found that many attenuated viruses alter their properties in different types of production cells. However, the focus here should be on documenting the properties of the virus and not the cell.

- How significant is short term testing of vaccines in large numbers of humans as it relates to the safety cell substrate?

In such requirements are we taking the regulatory role feared by Dr. Watson when he wrote of the suffocation brought about by the inaction of exercising unlimited prudence?

- How significant is the presence of a virus genome in a cell? Many countries produce rabies vaccines in such cells. In the U.S., we do not; is this unrealistic?

High yields of interferon can be obtained from certain transformed lymphoblasts. Should we reconsider the absolute contraindication to the use of transformed cells and view it more in a benefit vs. risk proposition that could vary with the circumstances? These and similar questions are what we expect you in the scientific community to come to grips with during this conference and in the future as we turn to you for guidance.

CELL SUBSTRATES: GENERAL OVERVIEW

Introductory Remarks

Saul Krugman, M.D.

Department of Pediatrics
New York University School of Medicine
New York, NY 10016

Dr. Harry Meyer has presented a superb overview of past achievements and future goals for the use of cell substrates in vaccines and other biologicals. In his closing comments he reminded us that when there was incomplete knowledge in the past, the formulation of certain rules and regulations was "dictated more by fears and uncertainties of the times than by reason".

The charge to the participants of this Symposium is clear. A critical review of the extensive experience accumulated during the past 25 years should provide the data needed for the assessment of the safety of various cell substrates. Have the potential risks proved to be real of spurious? The general overview of this first session will include a discussion of factors affecting the acceptibility of cell substrates, the philosophy of primary cells vs. cell lines and considerations of cells and safety.

CELL SUBSTRATES FOR BIOLOGICS PRODUCTION: FACTORS AFFECTING ACCEPTABILITY

John C. Petricciani, M.D.

Bureau of Biologics, Food and Drug Administration
8800 Rockville Pike
Bethesda, MD 20014

At the present time, the viral vaccines licensed for use in humans in the United States are produced in cells derived from a wide variety of sources. Table 1 shows the specific cell system for each vaccine and the year in which the vaccine was licensed. To some degree the cell type used for a given vaccine reflects the available technology at the time of licensure, and the then current consensus on what constituted an acceptable cell source. The classic example, of course, is poliovirus vaccine. Primary rhesus monkey kidney cultures were used initially. With the discovery of SV-40 as a rhesus agent capable of contaminating the vaccine, primary African green monkey kidney cell culture became the substrate of choice. More recently, diploid cells of human origin have been used to produce poliovirus vaccine, thus introducing an alternate acceptable cell substrate for vaccine production. Because new data have been developed at a rapid rate over the past decade in the biological sciences, and because there has been a large experience with cells from a variety of sources in vaccine production, it is of interest to examine which criteria are of major importance in assessing the acceptability of cells. The purpose of this discussion is to review those major criteria of acceptability which have been applied to cells used in biologics production up to the present time, and to critically analyze the logic and reasonableness of those criteria in the context of the general knowlege that is currently available in the biological sciences as well as the general experience with vaccines produced in a variety of cell types. The focus will be on cell lines rather than on both cell lines and primary cultures, since the use of primary cultures will be covered in depth by Dr. Parks, and I am sure that during the course of this symposium there will be many opportunities for us to discuss the issues relating to primary cell cultures.

Table 1

Human Virus Vaccines Currently Licensed in the United States

Vaccine	Year	Laboratory Host
Smallpox	1798	Human and bovine
Rabies	1885	Rabbit and embryonated duck egg
Yellow fever	1935	Mouse; embryonated chicken egg
Typhus	1938	Embryonated chicken eggs
Influenza	1940	Embryonated chicken eggs
Mumps (inactivated)	1945	Embryonated chicken eggs
Poliovirus (inactivated)	1955	Monkey kidney cells
Poliovirus (live)	1961	Monkey kidney cells
Measles (live)	1963	Chick embryo cells
Mumps (live)	1967	Chick embryo cells
Rubella (live)	1969	Duck embryo cells; rabbit kidney cells
Poliovirus (live)	1971	Human cell line WI-38

SOURCE OF CELLS

A generally conservative position was adopted early in the use of cells for vaccine production when it was decided that cells derived from normal tissues would be acceptable for use in manufacturing killed poliovirus vaccine, and that cells derived from neoplastic tissue would not be acceptable (1). The rationale for that position was reasonable since a virus vaccine produced in cancer cells could: (a) be contaminated with a hypothetical oncogenic agent associated with the neoplastic cells; (b) be contaminated with transforming nucleic acid of the cells themselves; or (c) have integrated the hypothetical neoplastic genes into the viral vaccine nucleic acid. These theoretical possibilities are illustrated in Figure 1. Primary cell cultures derived from normal tissues of normal animals became the acceptable system for use in vaccine production, and the specific cell type that was used for each vaccine depended on a number of factors, among the most important of which were: (a) viral susceptibility and yield; and (b) characteristics of an attenuated viral vaccine product in terms of ability to induce protective antibody without significant adverse side-effects in the recipients. The cell substrates which are in use today are those which have satisfied the above criteria best; the other substrates were discarded during the continuing evaulation of the vaccines. A recent example is rubella virus vaccine. Initially it was produced in three substrates: rabbit kidney, dog kidney, and WI-38. As experience grew, it became clear that the vaccine products were not equivalent with respect to side effects in children, and a shift in the choice of cell substrate began to take place which resulted in the elimination of the dog kidney substrate product from the market.

A different approach to the use of cells in vaccine production was made possible by the work of Swim and Parker (2) in 1957 and Tjio and Puck (3) in 1958 which demonstrated that cells could be grown from human tissues for a finite number of generations and that the cell cultures maintained a normal chromosomal pattern. Hayflick and Moorhead extended those studies in 1961 to serially propagate cells of fibroblast morphology from human embryonic lung, and showed that the cells maintained a normal diploid chromosomal constitution until they entered senescence (4). Such a cell system offered many potential advantages in that a standardized diploid cell line such as WI-38 could be used for many years with presumably predictable characteristics. But there were also objections raised regarding the use of WI-38 cells in human vaccines because of the theoretical possibility that they contained a human oncogenic agent for which no tests were available(5). In that regard, the objections were the same as those already mentioned for neoplastic cells.

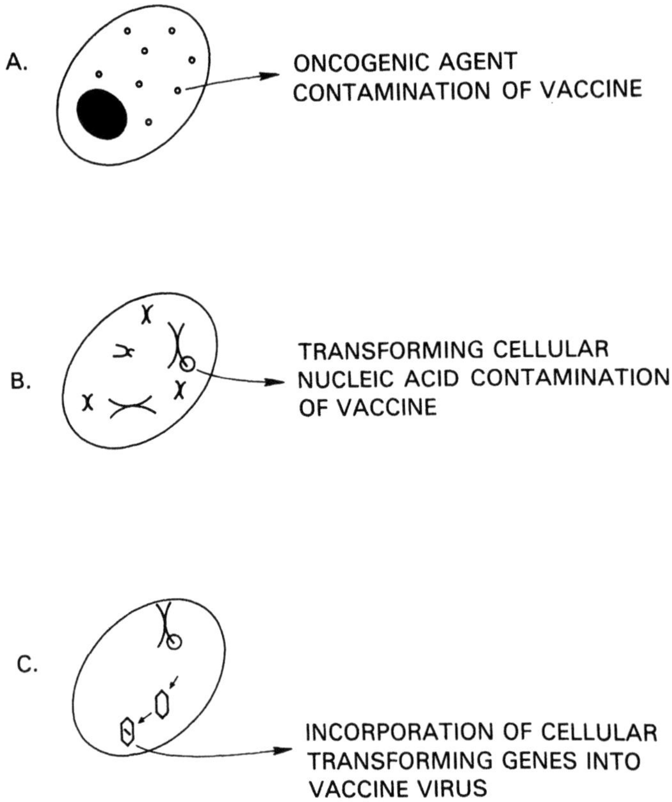

Figure 1. Theoretical Risks Associated with the Use of Neoplastic Cells in Biologics Production.

Nevertheless, there was gradual acceptance of the WI-38 diploid cell line in the production of live virus vaccines in Europe and eventually in the United States. Two of the studies which were incorporated into the total assessment of the acceptability of WI-38 were tests for tumorigenicity and karyology (6). It is important to understand the context within which these control procedures were devised. There was little or no experience with human diploid cells as substrates in vaccine production, and the cells were of human origin with the theoretical possibility that they contained an undetectable human tumor virus. Experimental arguments against those reservations at that point in time were extremely limited.

One could simply show that the cells themselves did not produce tumors in animal systems, and that they maintained a normal chromosomal pattern during their active growth period in vitro. It could be argued that even today we are not in a much better position in terms of offering solid experimental data to refute the proposition that human diploid cells contain an undetectable human tumor virus, since by definition such an agent would be undetectable. What can be said, though, is that no evidence exists to support that theoretical reservation even after every state of the art technique has been used to find such an agent.

The characteristics of other cells that possibly could be used as substrates in biologics production are contrasted with WI-38 in Table 2. Until recently, only cells with a finite life, normal karyology, and without tumorigenic properties have been seriously explored as substrates for use in biologics production, although each of the other known types of cells has found many uses in research and control testing. It should be pointed out, however, that human tumor cells (HeLa) were used for the production of an early experimental lot of live adenovirus vaccine without the development of tumors in the vaccine recipients 11½ years after inoculation (1). It is now about 22 years since that vaccine was administered to humans, and at least one of those individuals is alive and well and with us today. It would clearly be of interest to have health data on the other recipients. This example, in fact, illustrates two points which are central to the issue of which cells are, or are not, acceptable at any given point in time. First is the consideration of availability of alternative cell substrates to the one under consideration. The second point is the seriousness of the disease the product is intended to prevent. Those two factors are in turn interwined in the overall determination of benefits and risks associated with the product. If only HeLa cells could be used to cultivate poliovirus when there were almost 200,000 cases of polio reported in the United States for the 5-year period 1951-55, it would not have been unreasonable to attempt to balance the known risks of many people contracting poliomyelitis against the theoretical risks of using tumorigenic cells as the vaccine cell substrate. Because an alternative (monkey kidney cells) were available which appeared to be safer, HeLa cells were not approved for use. However, it is not always simply an issue of two or more alternatives that differ only in respect to potential risks. For example, if poliovirus had grown in monkey cells to a titer of only 10^3 $TCID_{50}$/ml for the production of the inactivated vaccine, one would have been forced to continue to consider HeLa cells simply because of the impracticality of using a cell system in which titers of only 10^3 could be obtained.

Table 2. Characteristics of Cell Lines Which Might Be Considered for Use in Biologics Production

	Life Span		Karyology		Tumorigenicity	
Example	Finite	Infinite	Normal	Abnormal	Positive	Negative
1. WI-38	X		X			X
2. Unknown at present	X		X		X	
3. Unknown at present		X	X		X	
4. HeLa		X		X	X	
5. BSC-1		X		X		X
6. Unkown at present		X	X			X
7. WI-38 (senescence)	X			X		X
8. Unknown at present	X			X	X	

KARYOLOGY

As mentioned earlier, particular emphasis was placed on the normal karyology of human diploid cell lines as an indication of their equivalence with normal cells and thereby their safety. A large pool of chromosomal data was accumulated, and standards of acceptability were established (7). Recommendations were also developed for safety tests during the production of vaccines in WI-38, and they included karyology monitoring of every lot of cells. Such monitoring in turn led to the development of an even larger body of cytogenetic data on WI-38 which tended to confirm the overall view that WI-38 cells displayed only low levels of chromosomal abnormalities until senescence. Because the use of human diploid cells to produce biologicals was new, and there was concern that some cultures might spontaneously transform, chromosomal monitoring was reasonable as a check on such an event since one would expect to see cytogenetic abnormalities accompanying the change to a neoplastic state. The large experience of many laboratories now strongly suggests that normal human cells do not transform spontaneously *in vitro* to neoplastic cells. Indeed, it has been extraordinarily difficult to transform human cells with chemical carcinogens, and there is still debate on the validity of most such reports which have appeared in the scientific literature. In view of the net experience with human cells in general, and WI-38 in particular, it would appear to be time to re-assess the need for the extensive chromosomal monitoring on each production lot of cells that has been required in the past. Such a review is in fact scheduled by an ad hoc committee which will meet after this symposium. Another reason which has been suggested for doing chromosomal analysis on each batch of cells is that it is useful in detecting cells which have been exposed to irradiation, certain chemicals, viruses, and mycoplasma infections (8). There is ample evidence in the literature that at least some mycoplasmas can induce chromosomal abnormalities (9), as can a number of other microbial agents such as viruses. It is also now well recognized that irradiation and some chemicals are potent clastogens. The pertinent question, however, is not whether or not various agents can induce chromosomal abnormalities; but whether or not karyology is a reasonable approach for the detection of extraneous biological agents in cells, or for the detection of cells that were exposed to physical or chemical agents. Karyology is simply not the method of choice for the detection of any of the known microbial agents, since specific tests are available for their isolation and identification, and, in fact, are required as part of the quality control procedures during the vaccine manufacturing process. To justify the use of karyology to detect cells exposed to physical or chemical clastogens, there should be some recognizable opportunity for such events to occur. By careful checks and controls on the procedures in the cell culture area and the media production area, such events can be virtually precluded, thus obviating the need for routine karyologic evaluation simply to detect such occurrences. It should be re-emphasized, however, that one can dismiss karyology for these purposes only if a reliable and functional control procedure is an integral part of the manufacturing process.

The reasonable focus of concern today should not be so much on repeatedly confirming the cytogenetic or other characteristics of WI-38 or other well-characterized diploid cell lines. Since those characteristics are now established and stable, we should focus rather on the identity of the cells being used in vaccine production.

Perhaps the single most important contribution of the diploid cell system is its potential in providing a well-characterized substrate <u>prior</u> to its use in biologics production. Is it not time now to take better advantage of this feature of diploid cells by relaxing the lot-by-lot monitoring, and instead focus attention on the identity of the cells to be sure they are in fact the well-characterized substrate they are supposed to be. Among the tests which can be used for cell identity is karyology. Therefore, karyology should be retained as a monitoring test, but the scope of that effort should be revised to meet this specific role.

Should serious consideration be given to cells other than primary cultures or diploid lines? There is, of course, no general answer that can be given to such a question since, as stated previously, a number of factors need to be balanced in arriving at a decision on any given case. However, there are several relevant points regarding the use of cells in vaccine production which are not well recognized and which should at least be surfaced in this presentation.

The use of cells which are chromosomally normal duing the first part of their life and which develop abnormalities as they age, and which may even go on to become established cell lines, has an enormous amount of precedence even though it may not be recognized as such. In the case of WI-38 we have, in fact, a system in which we use cells during the diploid phase of their life history even though it is now well recognized that in the later stages of their <u>in vitro</u> life chromosome abnormalities show a significant increase in frequency. Primary cultures of chicken embryo, duck embryo, rabbit kidney, dog kidney, and monkey kidney have been used to produce hundreds of millions of doses of live virus vaccines without regard to the eventual fate of those primary cultures in terms of showing that they maintained diploid characteristics or that they entered into a senescent phase and eventually died. Our experience as well as that of many other laboratories clearly shows that such primary cultures may not be diploid in the strict sense of the term, and that some may indeed go on to develop any number of chromosomal abnormalities (10). And at the same time, we have no indication that the use of such cell systems has imposed risks or hazards on the populations receiving the vaccines. On the contrary, the safety of vaccines produced in such cells has been so well established that we should be able to use that vast experience as a basis for at least the consideration of a much wider variety of cell lines in biologics production than simply diploid cell lines, or primary cultures if a special circumstance exists where the traditional alternatives appear to be inadequate.

SENESCENCE AND FINITE LIFE

In the same way that chromosomal abnormalities frequently are associated with at least the later passages of primary cultures, experience has shown that some will become continuous cell lines with chromosomal abnormalities. BSC-1 and LLCMK-2 are just two examples of such continuous cell lines derived from primary monkey kidney. And, of course, primary monkey kidney cultures are still being used in the production of poliovirus vaccine in the United States, Europe, and Asia. Thus, the fact of the matter is that for many years we have been using cells in vaccine production some of which have had the potential for an infinite life span. As mentioned previously, we have a huge experience with primary cell cultures and the safety of their use. Thus, it could be argued that the safety of using cells with an infinite life potential has been established at least in principal. The distinction between a cell line with a finite life and one with an infinite life may in fact be an artifact of our cell culture techniques, and this point will be discussed much more thoroughly by Dr. Kirkwood later.

TUMORIGENICITY

Tests for freedom from tumorigenic potential of cells used in vaccine production are currently required in the United States only to establish the original acceptability of human diploid cells. Tests on each batch of human cells for production are required by British Regulations. No National Control Authority requires such tests for any of the primary cell cultures. Again, a double standard appears to exist, the origin of which is found in the thinking that was prevalent in the 1960's regarding the safety of primary cultures and the potential risks of human diploid cells. Tumorigenicity tests of cell lines can be justified in the characterization of a new cell line which might be proposed for use in biologics production. However, routine tumorigenicity tests on each production batch of cells can be criticized on two counts. First, the only cells which have been shown to produce tumors in any of the currently available animal models are those which were: (a) derived from neoplastic tissue;

(b) transformed by an oncogenic agent; or (c) were derived from a species which carries a latent oncogenic virus. Since any new cell line will have been characterized regarding not only it tumorigenic potential but the presence of latent agents and chromosomal constitution, to repeat tumorigenicity tests on every batch of such cells would appear to be a redundant and futile exercise. The accumulated experience of many laboratories throughout the world confirms the fact that, with the exception of rodent cells, only those cells derived from neoplastic tissue or those transformed by an oncogenic agent have shown tumorigenic properties when tested. The second argument against the routine use of tumorigenicity tests relates to the sensitivity of the available tests and the basic questions that are being asked. If the question is whether or not the cell culture contains a mixture of "normal" and neoplastic cells, then tumorigenicity testing is probably not the method of choice to make that determination. For example, most current tests are able to detect about $5X10^5$ tumor cells. Thus, if the inoculum is 10^6 cells, then at least 50% would have to be tumor cells in order to be detected. Such gross contamination would be detected much more readily in even a screening type of chromosomal analysis than by tumorigenicity testing, if, as is usually the case, the tumor cells deviated significantly from the diploid state.

Tumorigenicity tests are extremely useful and appropriate in the characterization of a new cell line; but they are limited in their routine application because of sensitivity but more importantly because of the expense and the waste of animals and time in repeated testing of a cell line which is already known to be nontumorigenic.

CRITERIA OF ACCEPTABILITY BASED ON PRODUCT AND RISK ASSESSMENTS

The biological products with which we have had the greatest experience are, of course, the virus vaccines produced in cell culture systems. There are, however, two other types of biological products which should also be considered in any discussion of the factors affecting the acceptability of cells. First are the nonreplicating biological products which are synthesized by cells, and exemplified by interferon. The number of such protein by-products of cells is potentially very great. The second new type of cell culture product which is in an even earlier stage of development, is the cell culture itself. Specific functional cells such as insulin-producing islet cells grown <u>in vitro</u> are now being used experimentally in animals to treat diabetes. Thus, cell transplants appear as potentially important biological products in the future. The questions raised by each class of these new products is whether or not the same criteria of acceptability that now apply or may apply in the future to vaccine cell substrates also should be used for the new classes of products.

Biological products derived from cells can be divided into 3 general classes: (a) nonreplicating protein by-products synthesized by the cells; (b) microbial agents replicated in the cells; and (c) the cells themselves as transplantable producers of specific functional biochemicals. An examination of these 3 classes of products suggests that there is a clear difference in risk among them based solely on the consideration of the cell substrate involved. For example, the risk of using interferon produced from HeLa cells is probably much less than that of transplanting HeLa cells themselves into an immunosuppressed human. The reasons for the difference in risk relate directly to the nature of the final product plus the degree to which it can be purified and tested, in addition to the characteristics of the cells from which it was derived. The theoretical risks illustrated in Figure 1 become irrelevant to a cell product such as interferon if it can be shown that during the process of purification detectable nucleic acid is eliminated. Table 3 lists the general tests one might consider reasonable for the 3 types of cell products at the stage of characterization and in routine monitoring during biologics production. This scheme proposes that a cell line should be subjected to the same general studies regardless of the product class in order to establish a general data base against which later comparisons might be made once experience with the product becomes available. The difference in degree of testing is seen at the stage of routine monitoring where the amount of testing corresponds to the degree of risk (theoretical and real) associated with use of the the product.

Table 3

Cell Line Testing

Product Class	Characterization	Monitoring
Cell Transplants	Tumorigenicity, karyology, cell identity, latent agents, sterility, life potential	Tumorigenicity, cell identity, biochemical production, sterility, karyology
Replicating Microbial Agents	Tumorigenicity, karyology, cell identity, latent agents, sterility, life potential	Cell identity, sterility, limited karyology
Biochemicals	Tumorigenicity, karyology, cell identity, latent agents, sterility, life potential	Cell identity, Sterility

In summary, I have reviewed the factors which have played a major role in the assessment of acceptability of cell substrates in the past; and have attempted to draw attention to areas where there are inconsistencies in philosophy as well as in testing requirements when one compares primary cultures and cell lines. Unfortunately, we do not have as much knowledge as we would need in cell biology or virology to be able to resolve all of the questions which are associated with decision making in the choice of cell substrates. But that situation is not unique to cell substrates. In fact, it is frequently the case in biology and medicine that we are confronted with situations in which a choice must be made on the basis of rather incomplete information. Ideally, one should separate the scientific considerations from the emotional issues whether one is dealing with problems in areas as diverse as recombinant DNA or discussions on factors affecting the acceptability of cell substrates. This has not always been the case in the past. I hope that by looking at cell substrates from the points of view to be covered in this conference that we will be able to differentiate between science and emotion, and that at least some general principles will emerge regarding criteria for acceptability of cell substrates for the various biologicals currently available as well as for those in the developmental and research stages.

REFERENCES

1. Hilleman, M.R. Cells, vaccines and the pursuit of precedent. Nat. Cancer Inst. Monogr. 29:463-470. 1968.
2. Swim, H.E. and Parker, R.F. Culture characteristics of human fibroblasts propagated serially. Am. J. Hyg. 66:235-243. 1957.
3. Tjio, J.H. and Puck, T.T. Genetics of somatic mammalian cells II. Chromosomal constitution of cells in tissue culture. J. Exp. Med. 108:259-268. 1958.
4. Hayflick, L., and Moorhead, P.S. The serial cultivation of human diploid cell strains. Exp. Cell Res. 25:585-621. 1961.
5. Sabin, A. Discussion. Session V of the International Conference on Rubella Immunization. Amer. J. Dis. Child 118:378-381. 1969.
6. Minutes of the Third Meeting of the Committee on Cell Cultures. International Association of Microbiological Societies, Permanent Section on Microbiological Standardization, Pennsylvania, 18 May 1966.
7. Moorhead, P.S., Nichols, W.W., Perkins, F.T., and Hayflick, L. Standards of karyology for human diploid cells. J. of Biological Standardization 2:95-101. 1974.
8. Minutes of the Eighth Meeting of the Committee on Cell Cultures. International Association of Microbiological Societies, Permanent Section on Microbiological Standardization, Massachusetts, 4 October 1971.

9. Fogh, J., and Fogh, H. Chromosome changes in cell culture induced by mycoplasma infection. Annals of the New York Academy of Sciences, 225:311-329. 1973.

10. Wallace, R.E., Vasington, P.J., Petricciani, J.C., Hopps, H.E., Lorenz, D.E., and Kadanka, Z. Development and characterization of cell lines from subhuman primates. In Vitro 8:333-341. 1973.

QUANTITATION IN THE EVALUATION OF CELL SUBSTRATES FOR VIRAL VACCINE PRODUCTION

Wade P. Parks and Elizabeth S. Hubbell

Department of Microbiology, University of Miami
School of Medicine and Veterans Administration Hospital
Miami, Florida 33101

INTRODUCTION

Although the subject of cell substrates for viral production has recently been reviewed (1) certain points of considerable interest and relevance were not discussed. Issues of benefits and risks have been discussed in the present symposium in the preceding paper (2). In the present paper, we would like to stress the need for quantitative data related to genetic and presumably nongenetic alterations in various cell substrates. These changes may occur in the cells themselves or in their products. First, in order to discuss the issues of cell substrate selection certain definitions are necessary.

As applied to current vaccine technology, primary cells are usually monolayer cell cultures established from organs or embryos of animals which are dispersed mechanically with the aid of proteolytic enzymes before being plated and subsequently infected with the virus. At no time are the cells further subdivided and the virus growth occurs in cells which have undergone relatively few divisions. The most widely used example of a primary culture was the primary monkey kidney cells used for the production of attenuated live poliovirus vaccine or primary rabbit or duck cultures for rubella virus vaccine production. Cell strains are cells with a finite *in vitro* lifespan. For both primary cell cultures and cell strains, the more important criteria would be the number of cell divisions rather than passage level but for many reasons the former is used. The cell strain currently employed for vaccine production is the WI38 fetal lung strain used for the production of the live attenuated poliovirus vaccine. Cell lines are cell cultures which have acquired the ability to be propagated indefinitely *in vitro*.

Literally, cell lines have been selected for unlimited growth potential. A significant concern remains about cell lines as vaccine substrates because most such lines usually have been derived from tumors. As noted by Dr. Petricciani, in the preceding paper cell lines have also been derived from normal appearing cell cultures. Lymphoblastoid cell cultures which are derived from individuals without obvious clinical disease are cell lines by this definition. The increased growth potential in cell lines in many instances is clearly the result of the expression of known viral genes. For example most B-cell lymphoblastoid lines express one or more Epstein-Barr viral antigens (3).

The factors involved in the selection of cell substrates are rather obvious and have already been discussed (2). Risk factors related to cell substrate issues in current viral vaccines are generally low probability events (Table 1). Thus, in order to compare different cell substrates one must employ quantitation of low risk events. This may require not only _in vitro_ measurements but also _in vivo_ measurements. As an example of the extremes, immediate reactions in a recipient are fairly obvious if of a significant magnitude and specificity as in the case of poliomyelitis. In such a case, one can measure very low frequency events. Types of risks that are more difficult to evaluate are those that have a long latent period, less specific symptomotology, or occur in low frequency. We should also consider small subpopulations within the general population with unique susceptibility to untoward events related to viral vaccination above and beyond the obvious immunodeficiency diseases. Such abnormality would be very subtle and not easily detected with current techniques. However, this possibility remains to be fully evaluated and the possibility of either substrate-mediated vaccine viral alteration or some combination of the above must be considered.

Ultimately, the benefit/risk analysis of any given virus-cell interaction must be analyzed individually as a public health issue. This analysis must include an adequate evaluation of model populations and include prospective studies. However, in addition to the clinical sides of benefit/risk analysis, there are multiple events of every virus-cell substrate interaction that can be assessed and provide useful guides to selection of substrates for viral vaccines.

Current Primate Retroviruses

The current level of technology in viral vaccine manufacturing is probably good enough to avoid mass population problems. Nevertheless, it is clearly possible to impact significantly on vaccine acceptance by the general population in situations where relatively few individuals are directly harmed by untoward effects of viral vaccines. Administration of a viral vaccine contaminated with an

Table 1

Risk Considerations in Cell Substrate Selection

'Passenger Viruses'

1. Extrinsic - SV40, foamy viruses, avian

2. Intrinsic - retroviruses

Virus Alteration

1. Genetic - Adeno- SV40, erythroleukemia

2. Phenotypic - poliovirus virulence, rubella in dog kidney

Incorporation of host genetic material - recombination

unknown and/or oncogenic virus to mass populations remains as a continual concern of both manufacturers and regulators. Other major considerations related to "passenger virus" and vaccine virus alterations in cell substrates are listed in Table 1. Although steps to measure such potential agents exist currently in Federal regulations, they may not be adequate. The risks in terms of disease potential of such agents is not known but their mere presence could be enough to trigger national headlines.

No knowledgeable molecular biologist today doubts that man has tumor virus genes as a integral part of his genetic makeup. Many feel that expression of these genes occurs occasionally. This expression may be restricted to certain tissues as in the placenta (4); it may occur in certain diseases with increased frequency as in systemic lupus erythamatosus (5) or in certain genetic diseases (6,7).

Because of the work of Todaro, Beneveniste, Fine, Schlom and their associates, burgeoning information concerning primate retroviruses is available (8). Relatively few years ago there were no endogenous primate retroviruses, that is viruses which are inherited as part of the normal genetic makeup of monkeys. It is now clear that the normal cellular DNA both old world and new world primates contain vertically transmitted gene sequences of at least 3 type C and 2 type D retroviruses. These include the endogenous type C viruses of baboons, owl monkeys and stumptail macques and the 2 type D viruses of squirrel monkeys and langur monkeys (Table 2). These viruses are of particular interest because of their close evolutionary position to humans and the involvement of retroviruses in the genesis of naturally occurring cancer.

Table 2

Primate Retroviruses

Endogenous

Type-C Oncornavirus

1. Baboon
2. Stumptail macque
3. Owl monkey

Type-D

1. Squirrel monkey
2. Spectacled langur

'Exogenous'

Type-C Oncornavirus

1. Gibbon
2. Woolly monkey

Spumavirus

1. Cercopithecinae (Rhesus, Green)
2. Pan (chimp)

In each instance the isolation of transmissible retroviruses from primate cells has been a low probability event which requires extensive periods of cocultivation of primate cells with heterologous indicator cells which were permissive for viral replication (9). Once isolated it has been shown that all primates probably including man contain the genes for these viruses in their DNA (10). These viruses are present not just as a single gene but are represented as multiple copies in primate DNA. Each of the primate viruses are only distantly related to each other. Given the present rate of isolation of primate retroviruses and expanding knowledge of their biology it seems likely that as many as one in every one thousand cellular genes is a retrovirus gene (11). Although many of these virogenes may be defective, highly regulated, and expressed only infrequently, these sequences may be important not only for normal functions but also in the etiology of some human diseases.

The existence of integrated viral sequences is one of the major sources of concern of those who advocate nucleic acid-free vaccines or cell products from human or subhuman cell lines. An important point about the isolation attempts of endogenous retroviruses is that cell lines established from primates were among the first sources of these endogenous primate viruses. It should also be noted however that primary cultures from embryonic tissues have also yielded such viruses with comparable frequency (12). Granted, complete expression is an extremely low probability event in any case (13). Even if found in humans, the clinical significances of these sequences will be very difficult to establish.

Another major issue is the purely theoretical objection to these nucleic acids and considerations of their infectivity (14). Whereas man may receive infectious genes from human cell-grown virus vaccines there is 1) no evidence to date of the existence of such infectivity or 2) demonstrable clinical significance or 3) for the purposes of this symposium evidence that infectious nucleic acid sequences are any more common in cell lines than cell strains or primary cell cultures.

Recombination as a Basis of Viral Transformation

In the murine and avian systems it is now increasingly clear that the mechanism of transmissible viral transformation with retroviruses involves recombinational events. Examples in model systems where nucleic acid recombination has been implicated as the mechanism of tumorigenesis are shown in Table 3.

These studies in mouse and avian systems (15-17) answer part of the question "how" 'benign" viruses become capable of cell transformation. By virtue of recombination between viruses, viruses obtain the genetic information which endows them with the ability to increase the growth potential of the infected cell. The frequency of these recombinational events is extremely low, but it is possible using in vitro systems to select infrequent recombinants associated with cell transformation. Recombination without the acquisition of transforming ability probably occurs at a much higher rate (15). One should also recognize that at the present time, the measurement of transforming sequences both biochemically and biologically is technically easier than other viral genes.

The best example of natural recombination resolves a long standing question concerning the AKR mouse system. These mice express very high levels of virus 6 to 7 months before the development of thymic lymphomas that characterize this strain (16). The putative mechanism is that a highly specific albeit low frequency recombination occurs between two different endogenous murine leukemia viruses and then progeny virus is leukemogenic (16). Similarly, high multiplicity passage in vivo has resulted in the alteration of 'normal' leukemia viruses without oncogenic potential in animals to

Table 3

Current Endogenous Tumor Virus Systems which Demonstrate Recombination - 1978

AKR Thymic Leukemia

Friend Virus Erythroleukemia

Rous Sarcoma Virus

develop a highly virulent erythroleukemia as exemplified by the Friend leukemia virus (17). These recombinational events occur in a very specific fashion at the site of the evelope glycoprotein of one virus accompanied by a deletion of the nucleic acid from another virus. This recombinant virus is then a defective virus and requires a 'normal' nononcogenic replicating helper virus for infectivity in mammalian recombinants.

The liklihood of any risk to human patients if this model is operative in human cells would thus require the release of complete infectious virus. We know the infectious expression of human type-C viruses is extremely unusual, if it ever occurs (6, 13, 18).

The frequency of recombinational events cell lines and cell strains has been recently measured by Rapp and Todaro (19). They passaged nontumorigenic, nontransforming type-C viruses in normal and spontaneously transformed mouse cell lines and by careful selection have been able to isolate transforming viruses. As shown in the data in Table 4, the event was so very infrequent, that it cannot actually be measured but easily would appear to be less than one in one hundred million. Thus, the generation of transforming viruses from these types of cultures is a function of the cell substrate and the expression of specific genetic sequences. Since transforming viruses and the expression of different genetic sequences occurs in both 'normal' and tumorigenic lines it is obvious that simple karyology and/or tumorigenicity testing are inadequate to evaluate the likelihood of recombinational events.

Cell Line Variants

The most important issue of cell substrate evaluation is quantitation. In our studies of the regulation of transcription it became necessary to measure the frequency of change in cell cultures in the process of working with a given mammary tumor cell line. Over a course of several years, we frequently noted a spontaneous drop in mouse mammary tumor virus (MMTV) production in

Table 4

Frequency of Isolation of Transforming Sequences

Passage	Cell Morphology	No. Positive/ No. Attempts
0	Normal	0/12
1	Normal	0/10
3	Areas of criss-crossing	0/12
5	Criss-crossing plus rounding	3/7
11	Normal	0/8
20	Normal	0/12

From Rapp and Todaro, 1978

different mouse mammary tumor cell lines. This created several technical problems and led us to investigate in greater detail the specific mechanisms of this change (20).

The cell line used was first cloned to single cells using end-point dilution methods in microtiter plates. As shown in Table 5, sublines were tested and both the frequency of alteration of HGPRT expression (measured by resistance to thioguanine incorporation and cell death) and the levels of virus production were measured. Based on our knowledge of bacterial genetics one would predict a frequency of approximately 1 in a million cells spontaneously having an alteration in a single functional gene. In our studies, a frequency of about 1 and 500,000 cells was noted. This frequency was compatible with that of a single locus site, and further, in the presence of mutagens the frequency of conversion to resistance was increased approximately 10-fold. This suggested that these sublines from mouse mammary tumors behaved like other mammalian cell lines with regards to this genetic locus.

The frequency of high to low virus conversion was measured in the experiment shown in Table 6. Based on this single experiment and analysis of a total of several hundred individual clones, a conversion from high to low virus production occurs 1 to 2 times per 300 cell generations. This is an incredibly high frequency of variant formation. In nucleic acid hybridization studies it was shown that this alteration was at the level of transcription (21). In other experiments, individual clones were genetically stable once they became low expressors and no reversion has yet been detected. This process of high frequency variant formation and cell

Table 5

Alteration of a clone of C3HMT cells for growth in TG-containing media (10^{-5}M)

Clone	Treatment	Experiment No.	Frequency
296-14	None	1	2.9×10^{-5}
	None	2	5.6×10^{-5}
	None	3	2.6×10^{-5}
	None	4	7.1×10^{-5}
	None	4	5.0×10^{-4}

Table 6

Alteration of C3HMT cells with respect to expression of MuMTV polypeptides ($MuMTV^H$ -- $MuMTV^L$)

Experiment	$MuMTV^H$ clone designation[a]	$MuMTV^L$ clones/No. examined
1	296 Cl 14	2/73
1	341 Cl 101	2/77
2	396-7 Cl 2	3/30
3	296 Cl 143 TG^r[b]	1/32
4	296 Cl 143 TG^r	2/19

a Clonal populations were tested within 2 months of establishment from single-cell clones.

b TG^r = thioguanine resistant.

lines may or may not be unique to cell lines, but the marked change in phenotypic expression does emphasize the need for precise quantitation of phenomena in order to evaluate their origins and possible significance in vaccine substrate selection.

The phenomena of variant cell formation is not unique to the mammary tumor virus cell lines. Scharff and Coffino have extensively studied a mouse lymphoid line which produces heavy and light chain immunoglobulin (22). These cells frequently convert from the production of both heavy and light chains to the synthesis of the only light chain and also at a high frequency further convert to a cessation of extracellular immunoglobulin secretion (22). In these systems it would appear that variant cells may have some selective growth advantage but one must consider that the frequency of such alterations in cell expression may not be at all unusual. No obvious changes in morphology nor karyology were noted in the examples cited (20).

Concluding Remarks

This discussion has related primarily to low frequency events and their evaluation. Low frequency virus release, low frequency recombinations, selective pressures, and a somewhat higher frequency of variant cell formation provide some very early, even primative basis for quantitating cells as substrates for viral vaccines. By comparing the relative risks of these events with an even lower liklihood of all these events coming together in a human population, it seems unlikely they would be of detectable clinical significance. Nevertheless, issues concerning cell substrates or even general principles concerning cell substrates are at an early developmental stage. As further advances in cell biology are forthcoming, the significance of various quantifiable measurements may become clarified. In the meantime, decisions must be made on the basis of currently incomplete data. For example at present once relatively simple tests have been conducted there is no more (or less) hard evidence to support a decision to employ primary cultures or cell strains instead of human cell lines.

REFERENCES

1. Proceedings, Joint WHO/IABS. Symposium on the Standardization of Cell substrates for the Production of Virus Vaccines. S. Karger A.G. Basel. 1977.
2. Petricciani, J.C. Cell substrates in biologics production: Factors affecting acceptability. Preceding paper in symposium. 1978.

3. Klein, G., Pearson, G., Nadkarni, J.S., et al. Relation between Epstein-Barr viral and cell membrane immunofluorescence on Burkitt tumor cells. I. Dependence of cell membrane immunofluorescence on the presence of EB virus. J. Exp. Med. 128:1011-1020, 1968.
4. Benveniste, R.E., Lieber, M.M., Livingston, D.M. et al. Infectious type-C virus isolated from a baboon placenta. Nature 248:17-20, 1974.
5. Strand, M., August, J.T. Structural proteins of RNA tumor viruses on probes for viral gene expression. Cold Spring Harbor Symposium Quant. Biol. 39:pt 2:1109-1116, 1975.
6. Gallo, R.C. RNA Viruses, Genes, and Cancer in Genetics of Human Cancer. Edited by J.J. Mulvihill, R.W. Miller, J.F. Fraumeni, Jr. New York, Raven Press. 1977.
7. Huebner, R.J., Todaro, G.J. Oncogenes of RNA tumor viruses as determinants of cancer. Proc. Natl. Acad. Sci. 64:1087-1094, 1969.
8. Todaro, G.J, Sherr, C.J., Sen A. Endogenous new world primate type-C viruses isolated from an owl monkey (Aotus trivirgatus). Proc. Natl. Acad. Sci. 75:1004-1008, 1978.
9. Todaro, G.J., Benveniste, R.E., Sherwin, S.A. et al. MAC-1, a newly genetically transmitted type-C virus of primates: "low frequency" activation from stumptail monkey cell cultures. Cell 13:775-782, 1978.
10. Benveniste, R.E., Todaro, G.J. Evolution of type-C viral genes. I. Nucleic acid from baboon, type-C virus as a measure of divergence among primate species. Proc. Natl. Acad. Sci. 71:4513-4518, 1974.
11. Benveniste, R.E., Callahan, R., Sherr, C.J. et al. Two distinct endogenous type-C viruses isolated from the Asian rodent Mus cervicolor: conservation of virogene sequences in related rodent species. J. Virol. 21:849-862, 1977.
12. Todaro, G.J., Sherr, C.J., Benveniste, R.E. Baboons and their close relatives are unusual among primates in their ability to release nondefective endogenous type-C viruses. Virology 72: 278-262, 1976.
13. Stephenson, J.R., Aaronson, S.A. Endogenous type-C viral expression in primates. Nature 266:469-472, 1977.
14. Smotkin, D., Gianni, A.M., Rozenblatt, S. et al. Infectious viral DNA of murine leukemia virus. Proc. Natl. Acad. Sci. 72:4910-4913, 1975.
15. Wang, C.H., Duesberg, P., Beemon, U., et al. Mapping RN'ase T_1 - resistant oligonucleotides of avian tumor virus RNA's: sacoma-specific olignucleotides are near the poly(A) end and olignoucleotides common to sarcoma and non-transforming viruses are at the poly(A) end. J. Virol. 16:1051-1070.
16. Hartley, J.W., Wolford, N., Old, L.J., et al. A new class of murine leukemia virus associated with development of spontaneous lymphomas. Proc. Natl. Acad. Sci: 74:789-792, 1977.

17. Troxler, D.H., Boyers, J.U., Parks, W.P., et al. Friend strain of spleen focus-forming virus: a recombinant between mouse type-C ecotropic viral sequences and sequences related to xenotropic virus. J. Virol. 22:361-372, 1977.
18. Aaronson, S.A., Stephenson, J.R. Endogenous type-C RNA viruses of mammalian cells. Biochem. Biophys. Acta. 458:323-354, 1976.
19. Rapp, U.R., Todaro, G.J. Generation of oncogenic type-C viruses. Rapid leukemia viruses from C3H mouse cells. Science, In press, 1978.
20. Parks, W.P., Hubbell, E.S., Goldberg, R.J., et al: High frequency variation in mammary tumor expression in cell culture. Cell 8:87-93, 1976.
21. Coffino, P., Scharff, M. Rate of somatic mutation in immunoglobulin production of mouse myeloma cells. Proc. Natl. Acad. Sci. 68:219-223, 1971.
22. Young, H.A., Shih, T., Scolnick, E.M. et al. Steroid induction of mouse mammary tumor virus. Effect upon synthesis and degradation of viral RNA. J. Virol. 21:139-146, 1977.

HUMAN CELLS AND THE FINITE LIFESPAN THEORY

Thomas B.L. Kirkwood, M.A., M.Sc. and Robin Holliday, PhD.

National Institute for Biological Standards and Control, London NW3 6RB and National Institute for Medical Research, London NW7 1AA England

ABSTRACT

Cultures of human diploid fibroblasts are characterized by: i) finite lifespan, ii) marked heterogeneity in the growth potential of individual cells within the culture, iii) considerable variation in lifespans of parallel cultures of the same cell strain. To explain these properties, we have proposed a commitment theory of cellular aging. Cells are assumed initially to be uncommitted (potentially immortal) but, at each cell division, each daughter cell is assumed to have some fixed probability of becoming irreversibly committed to senesce and die after a specific number of cell generations. During the period between commitment and senescence, cells are assumed to multiply normally, so the uncommitted cells are diluted by committed ones and may be lost in subculturing. The theory explains features i) - iii) above and, in particular, suggests why diploid cultures have finite lifespans while transformed or permanent lines grow indefinitely. It also validly predicts the behaviour of mixed cultures of distinguishable but otherwise similar cell types, and that culture lifespan may be significantly decreased by drastic reduction of population size. The important converse prediction that culture lifespan may be extended indefinitely by growing sufficiently large cultures or by selectively isolating uncommitted cells remains to be tested.

INTRODUCTION

It is a well established characteristic of human diploid fibroblasts that they cannot be subcultured indefinitely (1-4).

Cultures derived from either foetal or adult tissue can sustain long periods of normal growth but these are invariably followed by a senescent phase and, subsequently, death of the whole population. Culture lifespan is inversely related to the age of the cell donor and is greatest for cultures derived from foetal tissue (3). Human fibroblasts which have been transformed by an oncogenic virus, such as SV40, or cells which have been derived from malignant tissue, such as the HeLa Line, can be grown indefinitely in culture. These permanent lines are heteroploid and have morphological and growth characteristics which are quite distinct from diploid strains.

To explain the finite lifespan of diploid cultures, it has sometimes been suggested that the cessation of cell division is due to 'terminal differentiation' and that it is controlled by a built-in genetic programme (5, 6). If this were true, however, it might be expected that, at any one time, all the cells in a population would be of roughly the same age, that is, at similar points along the programmed pathway from primary culture (phase I) through the period of steady growth (phase II) to senescence (phase III). This is clearly not the case, since it has been shown that individual cells taken from a population of fibroblasts are extremely variable in their doubling potential (5, 7, 8).

As an alternative explanation of the finite growth potential of diploid cultures we have proposed a commitment theory of fibroblast aging (9, 10). Starting with a population of uncommitted cells, which are potentially immortal, we assumed that there was a given probability that cell division will give rise to fibroblasts which are irreversibly committed to senescence and death. These cells initially multiply normally, but after a given number of cell divisions, which we call the incubation period, all the descendants of the original committed cell die out. We showed that if the probability of commitment was reasonably high and the incubation period sufficiently long, then the number of uncommitted cells in the population will progressively decline to <1 in 10^6, at which time, with normal subculturing regimens, they will inevitably be lost from the population. At this point all the remaining cells are committed to senescence and the population is therefore a mortal one, although different cells within it have very different growth potentials. We also showed that a reduction either in the probability of commitment or in the incubation period could produce an immortal steady state culture, consisting of a small subpopulation of uncommitted cells, a majority of committed cells and a constant fraction of nonviable cells. We suggested that transformed or permanent lines of mammalian cells may be in such a steady state.

Another important feature of diploid fibroblast growth, which again cannot readily be explained in terms of rigidly programmed senescence, is that there is considerable variation from one experiment to another in the actual lifespans of any one strain (11). We have shown that the stochastic nature of the process of dilution and loss of the uncommitted cells can satisfactorily account for this.

A number of specific predictions of the theory have been formulated and these have subsequently been shown to be in agreement with experimental results. This paper outlines the main features of the theory and its supporting evidence (for further details see refs. 10, 11, 12) and discusses its implications for tissue culture and related procedures.

CELL POPULATION STRUCTURE

Cells can be divided into three categories: uncommitted cells committed cells and dead cells. Any difference between dead cells and cells that have ceased to be able to divide is immaterial to discussion of population growth. (We assume that all such cells are transmitted, rather than lost, during routine subculture since it is known that many chemical or physical agents which kill cells do not prevent their attachment to solid surfaces. However, this assumption is not essential to the general conclusion we reach.) During the division of an uncommitted cell, we assume a probability P for each daughter cell that it becomes committed and a probability 1-P that it remains uncommitted. We thus have three possible outcomes to the division of an uncommitted cell:

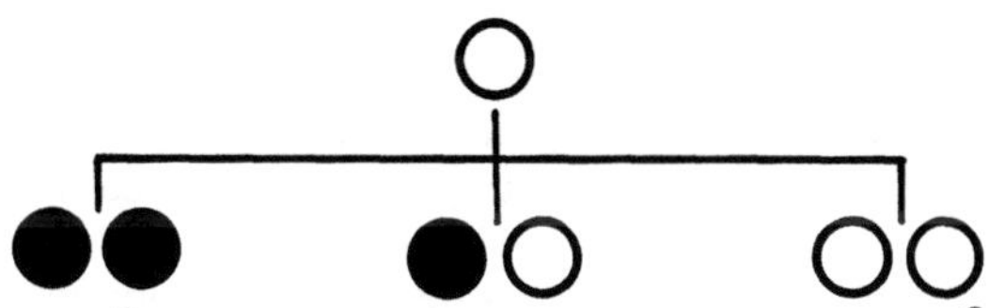

Probability: P^2 $2P(1-P)$ $(1-P)^2$

(○: uncommitted cell; ●: committed cell).

Commitment is assumed to be irreversible so that division of a committed cell always produces two committed cells. We assume that the path from commitment to death is the same for every cell and define the incubation period M as the number of cell divisions which elapse between commitment and death. (In defining M in this way we assume the cell cycle time to be constant, which is certainly an oversimplification. Variation in cell cycle time about a mean value has no radical effect on the general structure of the model.)

The committed cells can then be divided into M distinct subclasses, the first of which contains newly committed cells, the second of which contains cells that have divided exactly once since commitment, and so on. The last subclass contains cells that have divided exactly M-1 times since commitment and that will die after their next division.

The total population thus consists of M+2 classes, one of uncommitted cells, M of committed cells, and one of dead cells. The population structure is determined by the numbers of cells in each class, with transitions from one class to another occurring at each cell division.

We assume for the present that all live cells divide at the same rate. This assumption is probably unjustified, since there is evidence that cells divide more slowly as they approach death, but we make it to simplify the mathematical treatment. Slowing down the growth rate towards the end of the incubation period does not significantly alter the conclusions that we reach. It is also possible that committed and uncommitted cells might not divide at the same rate, in which case the rate of dilution of the uncommitted cells would be altered (12). In the absence of data on this, which must await the identification and isolation of uncommitted cells, we make the simplest assumption of equal division rates.

PREDICTED GROWTH CURVES

If we assume that a cell population is initially fully uncommitted and if we allow it to grow without restraint, we expect it to achieve one population doubling (PD) with each successive cell division until, at the (M+1)th division, cell deaths occur as the first committed cells reach the end of the incubation period. Inevitably the growth rate falls. What happens to the population then depends on the value of P. If $P > 0.5$, each uncommitted cell produces an average of less than one uncommitted daughter so that the uncommitted cells decline to extinction and the population dies out rapidly. If $P < 0.5$, the cell population growth rate falls to a level determined solely by the value of P and the population thereafter maintains this reduced but steady growth rate with a stable distribution over the classes. We term the first period of growth, prior to the first cell deaths, Stage 1, and the subsequent growth at reduced rate, Stage 2.

However, in the normal laboratory context populations do not grow without restraint. Populations of size N cells ($\simeq 10^6$) are repeatedly split (e.g. 1:2) and the subcultures allowed to grow until they again contain approximately N cells each. The essential feature of this procedure is that at any time no single unit of the population contains more than N cells. During Stage 1 growth the accumulation of committed cells dilutes the uncommitted cells even though the potential total number of uncommited cells may be increasing (for $P < 0.5$). There may thus come a point where the uncommitted cells are so dilute that there is a high probability that individual subcultures contain no uncommitted cells, and are therefore mortal. Alternatively, if a reasonable number of uncommitted cells remain in the individual subcultures at the transition from Stage 1 to Stage 2, the adoption of a stable distribution of cells over the classes means that no further dilution of the uncommitted cells occurs. Thus, under these circumstances, cultures may continue to grow indefinitely.

For this reason both the population size N and the incubation period M play important parts in determining the mortality or immortality of the population. For a given value of P, if N is too small or M is too large the uncommitted cells are likely to be lost during Stage 1 and the populations are then certain to be mortal. When this happens, the growth rate will still show the transition from Stage 1 to Stage 2 growth as before and will remain steady for a number of cell divisions in Stage 2. However, as the last cells to become committed approach death, the growth rate will again begin to fall, this time steadily to extinction. We term this final stage of growth Stage 3. The transition from Stage 2 to Stage 3 is not a sharply defined event but corresponds roughly with the appearance of visible senescence in the culture.

The general predictions of the model are thus as follows. For $P > 0.5$ cultures will grow for a time at a steady rate and will then show a rapid decline followed by death. For $P < 0.5$, cultures will grow at a steady rate until the first cell deaths occur when they change to a reduced, but again steady, growth rate. Depending on the combination of values of P, M and N the cultures will then either senesce and die or will grow on indefinitely (Figure 1). The predicted sudden change in growth rate at the transition from Stage 1 to Stage 2 is specific to the model and it is interesting that such an effect has been observed by us and by others (J.P. Jacobs, unpublished observations). Indeed a growth experiment performed using MRC-5 fibroblasts (13) by one of us (RH) some years ago gives a close fit to the model's prediction (10).

HETEROGENEITY WITHIN CELL POPULATIONS

With each generation, cells transfer from one subclass of the population to another. An ever-diminishing pool of uncommitted

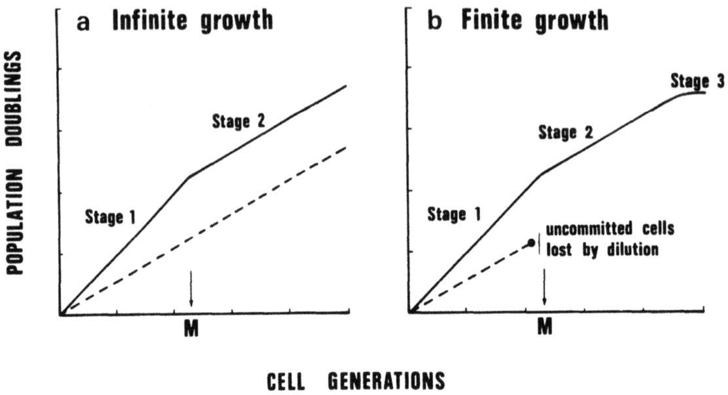

Figure 1

Alternative cell culture growth curves predicted by the commitment theory of cellular aging for an initially fully uncommitted cell population. The total cell population (solid line) doubles with each cell generation (Stage 1) during the first M (incubation period) generations. The uncommitted cells (broken line) multiply at a lower rate since a fraction P (probability of commitment) become committed at each division. After M cell generations some cells die and the growth rate falls, stabilizing rapidly to a steady rate of cell multiplication by a factor 2(1-P) per cell generation (Stage 2). a) The uncommitted cells survive the progressive dilution during Stage 1 to remain present in Stage 2. The culture grows indefinitely with a stable population structure. b) The uncommitted cells are lost by dilution (see text) during Stage 1 and the culture is then doomed to mortality. After a strictly limited period of growth in Stage 2, the culture senesces and dies (Stage 3).

cells generates new committed cells until finally (in the case of diploid cultures) it is lost altogether by dilution. Meanwhile, each new cohort of committed cells works its way steadily through the incubation period, losing some of its division potential with every additional step. Ultimately, it reaches the end of its incubation period and forms part of an accumulation of dead cells.

At any point during a culture lifespan, the population is therefore a heterogeneouus collection of cells of differing division potentials and its structure can be presented as a 'profile' of the numbers of cells in each subclass. Figure 2 illustrates the profiles of a typical culture at four different points in its growth. We must note that laboratory populations of fibroblasts have always gone through many cell generations in vivo and during primary (Phase I) growth in vitro so the early profiles in the simulated growth of an initially uncommitted cell population will not be seen in practice.

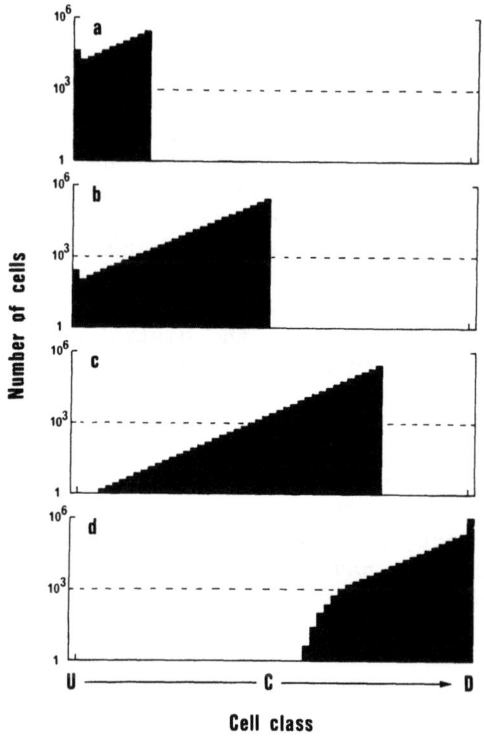

Figure 2

Theoretical population profiles, that is, distributions of cells in the classes of the model, at four successive points during a culture lifespan. Uncommitted cells (U) give rise to committed daughter cells, which move steadily to the right through the committed classes (C) until at the end of their incubation period, they die. Profiles a, b, c belong to Stage 1 when all cells are still alive; profile d belongs to Stage 2 when dead cells are present.

The heterogeneity of division potential within the population at later times (e.g. Figure 2d) is broadly similar to what has been observed (5, 7, 8). Detailed comparison of the theoretical and experimental results is hampered at present by the limited realism of our assumption that committed cells do not divide more slowly as they approach death. We are currently working towards modification of the theoretical model to accommodate a more realistic form of this assumption. This will alter the shape of the right-hand end of the profile in Figure 2d.

VARIATION IN CULTURE LIFESPAN

So far, we have taken no account of the stochastic nature of the process of dilution and loss of the uncommitted cells. In reality, there is randomness in the way cells become committed, in the selection of cells during subculturing and in the subsequent division of the cells that make up the subculture. We have shown, by incorporating these random elements into a stochastic model, that the theory predicts considerable variation between parallel cultures in the actual times of loss of the uncommitted cells and that this, in turn, causes significant variation in the lifespans of cultures (11, 12). Comparison of the distribution of lifespan of 300 simulated cultures with 67 independent populations of MRC-5 fibroblasts (Figure 3) showed a reasonable measure of agreement between theory and experiment.

It is perhaps surprising that variation in the time of loss of the small remaining subpopulation of uncommitted cells should have such a marked effect on culture lifespan, but it must be remembered that these cells have considerably greater division potential than the vast majority of the population.

EFFECT OF CULTURE SIZE ON LIFESPAN

One of the most important predictions of the theory is that the number of cells in a culture is a major factor in determining the longevity of fibroblasts, since the population size is directly related to the probability of losing the last uncommited cells from the culture. For example, the smaller the culture the less time it will take to dilute out the uncommitted cells and the shorter will be the lifespan.

This prediction was tested experimentally with MRC-5 cells. Cultures of normal size ($\simeq 10^6$ cells) were drastically reduced or 'bottlenecked' to about 10^3 cells at selected times during

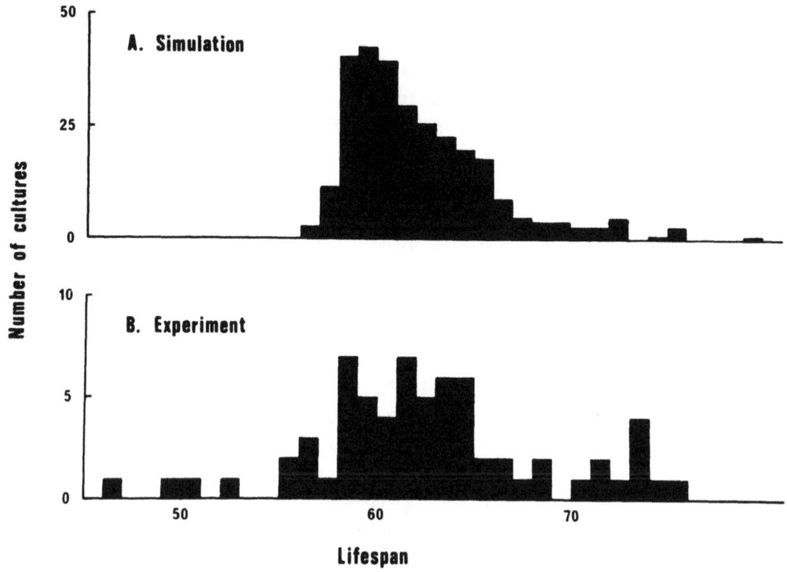

Figure 3

Comparison of the distribution of lifespan of 300 simulated diploid fibroblast cultures (P=0.275, M=55, N=2x10^6 with 67 independent populations of MRC-5 fibroblasts (for details see ref. 11). The experimental lifespans are measured in population doublings (PD's) from establishment of the primary culture; the simulated lifespans are expressed on a similar scale but with indeterminate origin since we cannot at present determine how many fibroblast generations may have elapsed in vivo and in vitro prior to establishment of the primary culture. (For the above parameter values the mean simulated lifespan was 77 PD's from the time the population was fully uncommitted.)

their growth and were then grown back to their original size and cultured until senescence. According to our calculations, this bottlenecking was expected to cause a significant reduction (≃ 8PD's) in mean culture lifespan when performed within a wide range of PD or passage levels. The results of an extensive experiment almost exactly confirmed this (11).

GROWTH OF MIXED CULTURES

When a culture consisting of a mixture of two distinguishable but otherwise similar cell types is grown, it would normally be predicted that the proportions of cells of the two types should remain more or less constant, perhaps with a slight drift if one or other type had a minor advantage. However, the commitment theory predicts that, towards the end of their lifespan a high proportion of such cultures should show pronounced tendencies to a predominance of either one or the other cell type due to differences in the time of loss of their respective uncommitted cells.

Unpublished data of C. Zavala, P.J. Fialkow and G. Herner (cited in ref. 11) support this prediction. They established skin fibroblast cultures from female donors who were heterozygous for the allelic variants A and B of glucose-6-phosphate dehydrogenase. Since G6PD is X linked and only one X chromosome is active, individual cells have either variant A or B. Most cultures initially contained approximately equal numbers of each cell type, and in the most cases the A/B ratio stayed constant for many PD's indicating that there was no selection for one cell type. However, at the end of the lifespan it was frequently found that either A or B cells became predominant.

CONCLUSIONS

In conclusion, the commitment theory of fibroblast aging explains: (i) The finite lifespan of diploid fibroblast cultures, (ii) The heterogenity in cell division potential within the cultures, (iii) The variation in lifespan between parallel cell cultures. In addition, it makes three specific predictions which have been verified by experiment: (iv) The fall in growth rate corresponding to the transition from Stage 1 growth to Stage 2, (v) The effect on culture lifespan of changing population size, (vi) The growth behaviour of mixed cell populations.

So far as we are aware, no other single explanation can account for all these phenomena.

A scientific theory is of merit only if it provides a unified basis for interpreting already known facts and if it generates experimentally testable predictions so that it can, in principle, be falsified. In terms of existing data, the commitment theory underlines the experimental observations on clonal heterogeneity within fibroblast population and fits, in general, with the recent observations by Smith *et al.*(8) on the changing pattern of this heterogeneity with age. As Smith and his colleagues showed, this property may be used as an indicator of both *in vitro* and *in vivo* human cell aging. The theory also provides an interpretation of the variability in longevity of cell cultures of the same strain. This is a factor which should be taken into account when, for example, characterizing the longevity of new cell strains or performing diagnostic tests for diseases of premature aging. Reliable estimation of lifespan depends on the serial subculture of several independent populations. In addition, the bottleneck experiment shows that large changes in population size, even only temporary ones, can significantly alter culture lifespan.

Of the further predictions to be tested the most important is that, if the theory is valid, fibroblasts are not intrinsically mortal and the Hayflick limit to their lifespan is, in a sense, an artifact that is the inevitable consequence of the normal laboratory procedures for tissue culture. The theory predicts that if a sufficiently large culture is established early on, when uncommitted cells are still present, it should be possible to avoid the loss of the uncommitted cells and to maintain the population in steady growth indefinitely. On the basis of our best current estimates for the probability of commitment (0.275) and for the incubation period (approximately 60 cell generations), we calculate that 10^{10} cells may be enough (these estimates are derived from results with MRC-5 fibroblasts and may be slightly different for other cell strains). This is a large number but it is not outside the capacity of a well equipped laboratory, and the long term benefits that would result from being able to maintain indefinitely a standardized steady state diploid cell population are sufficient to justify research in this direction. As an alternative to growing very large cultures, it may prove possible selectively to isolate uncommitted cells or cell populations enriched with uncommitted cells. This would also allow indefinte propagation of the diploid cell strains.

Finally, the commitment theory may also help to throw light on the fundamental difference between mortal diploid cells and immortal heteroploid ones since reduction in the incubation period or the probability of commitment can convert a population with finite growth into one with infinite growth potential. If this is the case, then both types of population contain large numbers of cells committed to a finite division potential. The difference between them is that the transformed cell populations also contain stable subpopulations of immortal uncommitted cells.

REFERENCES

1. Hayflick, L., Moorhead, P.S. The serial subcultivation of human diploid cell strains. Exp. Cell Res. 25:585-621, 1961.
2. Hayflick, L. The limited *in vitro* lifetime of human diploid cell strains. Exp. Cell Res. 37:614-636, 1965.
3. Martin, G.M., Sprague, C.A., Epstein C. Replicative life-span of cultivated human cells. Lab Invest. 23:86-92, 1970.
4. Thompson, K.V.A. and Holliday, R. Effect of temperature on the longevity of human fibroblasts in culture. Exp. Cell Res. 80:354-360, 1973.
5. Martin, G.M., Sprague, C.A., Norwood, T.H., *et al*. Clonal selection, attenuation and differentiation in an *in vitro* model of hyperplasia. Am. J. Pathol. 74:137-153, 1974.
6. Hayflick, L. The biology of human aging. Am. J. Med. Sci. 265:433-455, 1973.
7. Smith, J.R., Hayflick, L. Variation in the life-span of clones derived from human diploid cell strains. J. Cell Biol. 62: 48-53. 1974.
8. Smith, J.R., Pereira-Smith, O.M., Schneider, E.L. Colony size distributions as a measure of *in vivo* and *in vitro* aging. Proc. Natl. Acad. Sci. USA 75:1353-1356, 1978.
9. Holliday, R. Growth and death of diploid and transformed human fibroblasts. Fed. Proc. Fed. Am. Soc. Exp. Biol. 34:51-55, 1974.
10. Kirkwood, T.B.L., Holliday R. Commitment to senescence. A model for the finite and infinite growth of diploid and transformed human fibroblasts in culture. J. Theoret Biol. 53:481-496, 1975.
11. Holliday, R., Huschtscha, L.I., Tarrant, G.M., *et al*. Testing the commitment theory of cellular aging. Science 198:366-372, 1977.
12. Kirkwood, T.B.L., Holliday, R. A stochastic model for the commitment of human cells to senescence. Biomathematics and Cell Kinetics. Edited by A.J. Valleron, P.D.M. Macdonald. Elsevier/North-Holland, Biomedical Press, 1978, pp. 161-172.
13. Jacobs J.P., Jones, C.M., Baille, J.P. Characteristics of a human diploid cell designated MRC-5. Nature 227:168-170, 1970.

LINE CELL SAGA - AN ARGUMENT IN FAVOR OF PRODUCTION OF BIOLOGICS IN CANCER CELLS

Maurice R. Hilleman, Ph.D., D.Sc.

Division of Virus & Cell Biology Research
Merck Institute for Therapeutic Research
West Point, Pennsylvania 19486

INTRODUCTION

William Shakespeare observed that "All the world's a stage and all the men and women merely players!" This is no less true for those who act out the events of science than for thespians who pursue a career in the theater. It is amply illustrated in the drama that has attended the evolution of the dogma for acceptability or nonacceptability of particular kinds of cell cultures for preparing biologic products for use in man. Even cursory review leads to the conclusion that ill-conceived fears, preconceived notions, and personalities have contributed as much, perhaps, to decision-making as has science itself in this area of activity in which the scientific fact base is less than definitive.

The epicenter for discussions of cell acceptability has been that of safety or perceptions of safety, principally as they relate to possible induction of cancer. The dogmas that were created initially were strongly forged and have been revised only in minor degree to the present time. The problem is that economic factors were not a part of the equations that guided decision making and this has created a roadblock of substantial proportion to future production, in the practical sense, of new products that demand mass synthesis of the active substances whether they be viruses or other products of cells. To modify and to correct is best assisted by understanding, and it seems worthwhile to review those past events that have provided the historic precedents for contemporary rules.

ORIGINS OF SPURIOUS CONCEPTS

It all began on a cold and dismal January 22, 1954 (1-3) at the Walter Reed Army Institute of Research. Dr. J. Werner and I (4) had discovered a new family of viruses, now called adenoviruses, that we found to cause primary atypical pneumonia (PAP) and acute respiratory disease (ARD) in military recruits. We had linked these viruses serologically and epidemiologically to human disease but two questions remained. Was there need for human volunteer experiments to complete fulfillment of Koch's postulates? And what about vaccines? Dr. John Dingle of Western Reserve was prepared to do the human volunteer tests. We were prepared to develop a vaccine. But the hang-up on both questions was with cells. Thus far, the viruses grew well only in HeLa cells and these originated from a human cancer. Could materials of HeLa cell origin be used in man?

Twenty-five scientists of diverse discipline braved the storm of January 22 to attend the meeting that had been called by the Armed Forces Epidemiological Board to consider the question. The meeting lasted for only two hours and the declaration was feeble, but the pronouncement initiated a precedent that has been expanded and extended to form current policy. It was simply stated: "Virus material grown on normal cells inactivated by formalin is acceptable for limited immunization trials." Thereby, cell culture originating from frankly neoplastic tissues or possessing attributes of neoplasia were definitely ruled out. And we "had had it" insofar as HeLa cell vaccine was concerned. Normalcy of cells was not defined, and the subject of continual passage of cells of "normal" tissue origin was avoided. Dr. Dingle's fulfillment of Koch's postulates was deemed unnecessary. It should be noted that these decisions of a quarter century ago were made at a time when very little was known about tumor immunology and transplantation rejection, and the real perceived fear in the use of HeLa and other transformed cells was possible inadvertent transfer of intact cells and of cancer.

The question of continuous cell lines for vaccine preparation was the subject of a second conference (1,2) that was convened by the Armed Forces Epidemiological Board in Atlantic City, New Jersey, on April 12, 1959. The diversity of opinion expressed at this second meeting evidently precluded any definitive conclusion but cancer was still the main concern, and it was made clear that the possible risk attending use of continuous cell lines would be regarded as too great for vaccines against diseases of only "mediocre" importance.

This view eventually became a rule stated in the 1967 U.S. Public Health Service Regulations for Biologic Products (5) that forbade use of continuous cell lines for manufacture of live poliovirus vaccine and for live and killed measles virus vaccines. Such regulation was not uniformly accepted and became the subject of a major controversy (6-16) that is familiar to most workers in the field.

The first significant event in the move toward formulation of official regulatory policy was the appointment in 1962 by the Director of the National Institutes of Health of a Committee on Tissue Culture Viruses and Vaccines comprised of staff members of the National Institutes of Health. The Committee report, published in 1963 (7), was rambling and verbose, and was more noteworthy for its recognition of possible problems than for its providing solutions. Like so many bureaucratic reports, it didn't say yes; it didn't say no; it just called for more studies.

Essentially, the Committee took cognizance of the existence of three basic kinds of cell cultures used to prepare viral products. These were: (a) primary cultures of cells freshly explanted from normal tissues, (b) continuously cultured mixed cell populations derived from normal tissues, and (c) stabilized cell lines that were continuously cultured. The Committee recognized the known problems of primary cell cultures, especially those relating to presence of indigenous viruses that in some instances were frankly oncogenic for cells in culture or for subhuman animal species. The Committee was vocal in stating the advantages of continuously cultivated cell cultures and these included independence from a continual supply of fresh tissue, greater assurance of freedom from adventitious viruses harbored by the primary tissue, and the means for accumulating and storing certified cell seed stocks. The added advantages of more economic cell production and the possibilty for cell propagation in chemically defined media free of exogenous microbial contaminants was also noted. In opposition, the Committee went beyond the original concern for cancer transplantation and raised theoretical questions as to whether the cell-free products of serially passaged cells might contain some hypothetical virus, nucleic acid, or nucleoprotein that might be infective and pathogenic *in vitro* or *in vivo* under some conditions for some species.

CONFLICT AND CONTROVERSY

The report might have gone quietly by as just one more opus created by a committee had it not contained a statement that tread squarely on the toes of Dr. Leonard Hayflick and his associates (6,8). These workers had developed a number of cell strains from human embryonic tissues and they had quite different views on the matter. The offending statement in the Committee report was that "...continuously cultured cells eventually develop characteristics suggestive of malignant change" and it was referring to the neoplastic alteration accompanied by heteroploidy that had been observed for certain kinds of animal cells in culture. Hayflick pointed out that such alteration was indeed a rare event with normal human cells, especially since cells derived from normal human tissues typically displayed finite ability for replication, perhaps up to 50 passages, with termination thereafter. Most importantly, the Committee failed to recognize the difference between the unaltered diploid cell strains and altered heteroploid cell lines. Failing this distinction, Hayflick's diploid cell would, by definition, be excluded from vaccine use.

Dissatisfied with U.S. regulatory action, the forum was moved to the Committee on Cell Cultures of the Permanent Section on Microbiological Standardization of the International Association of Microbiological Societies (World Health Organization affiliated) under the leadership of Drs. F. Perkins and L. Hayflick. In its report of 1967 (12, see also 17, 18), this Committee rejected outright the concept that only primary cell cultures could be used for virus vaccine production. Instead, it recommended that serially propagated human diploid cells be considered a satisfactory substrate for production of human viral vaccines provided that several criteria were met covering (a) derivation from normal human fetal tissue, (b) lack of significant deviation in karyology from specified norms, (c) presence and retention of normal morphologic and biologic characteristics, (d) absence of extraneous microorganisms, (e) lack of *in vivo* tumorigenicity in animals, (f) sufficient supply to be practically useful, (g) capability of giving satisfactorily high yields of virus, and (h) approval by the National Control Authority. It was recorded that more than 200,000 persons had already received vaccines by both parenteral and oral routes that were prepared in Hayflick's WI-38 diploid cell strain and that there were no untoward effects.

Following such definition and leadership, the diploid cell stain concept was vigorously pursued in East and West Europe and approvals were granted by several National Health Authorities for use of vaccines produced in WI-38 cells. Opposition was continued in the USA, now based mostly on the concept put forth by Sabin (9-11) of the possible presence of human leukemia virus in diploid cell strains. No real basis was found to support this speculation. Meantime, storm clouds were gathering over live poliovirus vaccine produced in primary cell cultures (see 2). These involved increasing problems with indigenous viruses of monkey kidney including frankly oncogenic SV_{40} virus (19), the threatening diminished supply of monkeys, and the deaths that occurred in a number of persons who came into direct or indirect contact with grivet monkeys originating in Africa that were infected with the so-called Marburg virus (20). In March 1972, the U.S. regulatory agency believed that the questions relating to safety of products made in WI-38 were satisfactorily resolved and live oral poliovaccine prepared from virus grown in WI-38 cells was licensed for general use. In March 1977, a license was given for a rubella vaccine produced in WI-38 cells and administered parenterally.

THE NEED FOR CHANGE

The decision to license vaccine produced using virus grown in diploid human cells marked a notable advance in vaccine technology, making possible the resolution of live poliovirus vaccine problems. At the same time, it opened the door to new vaccines against such agents as respiratory syncytial virus, varicella, and cytomegalovirus for which diploid human cells provide the only readily available and acceptable means for propagation.

Having proceeded this far with serially propagable cells, the proposition might be made for extending policy to cover all serially propagable cells, including those of frank neoplastic origin such as HeLa or KB cells. The question of why might reasonably be asked and the answer would be mainly economics - to bring costs within the possibility for fulfilling human needs.

Any serious consideration for use of cancer cells for vaccine purpose might be simplified from the start if all considerations be limited to nonliving biologics alone - excluding all live virus products for which safety might be more difficult to establish. There is, to my knowledge, no present important live virus vaccine need for which primary cells or diploid cell strains will not provide the acceptable answer. To the contrary, with killed vaccines, and especially subunit vaccines containing only the immunizing glycoprotein and polypeptide components of the virus, the problem is very difficult since a comparatively large amount of viral substance is required for each dose of vaccine, and this is why cells propagable in tank culture must be utilized. The same need for mass production of low yielding substance supplies also to the preparation of exogenous interferon for human administration. For such products, submerged cell culture offers basically the same technologic advance as conversion from bottle to tank culture did in the early days of penicillin production.

There are four critically important disease problems of the contemporary era that, alone, could justify the effort that will be needed to supply the data base for eventual licensure of cancer cells in production. The first of these is killed poliovirus vaccine that presently is made in primary monkey kidney or in diploid cells at considerably greater cost than can be paid by the nations that need them most. Live poliovirus vaccines have failed in tropical climates, largely because of interference brought about by intercurrent infection with other viruses that prevent a vaccine take in a substantial proportion of the population (21). Killed vaccine does not have this problem, but present costs render it prohibitive in economically developing societies.

A second problem is that of genital herpes simplex that has become increasingly important with changes in social mores. The infection, because of its recurrent nature, is now regarded by many public health workers as the number one venereal disease. Viral propagation can be achieved in primary chick embryo cells and is just within the bounds of economic reality but only for affluent populations that can afford fairly costly procedures. Propagation in line cells in tank culture might reduce costs very substantially.

Hepatitis B vaccine (22) is a third product candidate that is well along on its development path. The virus fails to replicate significantly in any kind of cell culture in the laboratory, and the only source of antigen for vaccine preparation is from human beings who are persistently infected with the agent and who circulate large amounts of the surface antigen of the virus in their blood. The relatively high cost for infected plasma and the costs for processing are serious deterrents to low-cost manufacture of the product. During 1976, Alexander and associates (23) reported recovery of a liver cell line from a case of human hepatoma that proliferated well in cell culture and that continuously shed at least part of the surface antigen complex of hepatitis B virus in economically important amounts. Utilization of the Alexander cell as a source of hepatitis antigen for vaccine use might be optimistically viewed if frank neoplastic cells can be certified for this purpose.

Fourth, there is the current resurgence of interest in the application of interferon (24-26) to the therapy of certain human illnesses of grave consequences including acute and persistent infections with viruses of the herpesvirus family, chronic hepatitis B, and certain kinds of cancer in man. There is no economically acceptable source of interferon at the present time. Production of interferon in freshly collected human leukocytes gives satisfactory products for research purpose but hardly qualifies for mass production at a reasonable cost. Production in diploid fibroblast cells in culture gives a product of low level stability and effectiveness in man and is of limited promise (24). Production in lymphoblastoid cell lines such as the Namalva cell (27, 28) derived from tumor tissue from a case of Burkitt lymphoma, to the contrary, shows great promise as a reliable and economically acceptable source of interferon but its neoplastic origin precludes its consideration for other than experimental use at the present time. Such lymphoblastoid cell lines represent B cell lymphocytes that were transformed by the virus of infectious mononucleosis (Epstein-Barr virus) and that carry genetic pieces of the virus in differing numbers and sites both as DNA integrated into the chromosome of the cell and as plasmid-like structures (29-31). Cell lines, such as the Namalva cell, are not productive of infective virus. Such integration into the cell genome has also been demonstrated for adenoviruses (32) and SV40 virus (33) that also are oncogenic.

The problem of application of line cells, especially of neoplastic cells, to biologics production is clearly a problem whose time has come. In seeking resolution, the most difficult problem is to define what is the problem in the first place since the interface is with shadow fantasies and fears that have no real data base to support them and that hardly even deserve being accorded the distinction of a plausible theory. The fact is that while something is known about the events that surround human cancer, the primary cause remains unknown. More important to the biologics problems, it is unknown whether the disease is transmissible in any case.

RATIONAL RESOLUTION

Within the framework of this discussion, it may be observed that most ideas about cancer etiology center on two basic concepts that one may call the <u>exogenous</u> school and the <u>endogenous</u> school. The <u>exogenous</u> school, led by virologists who study cancers in animals caused by viruses, see the placement into the cell of new genetic information as the cause of cancer. These workers favor elimination of primary cell culture from biologics use in which the viral "flora or fauna" cannot be defined and detected with certainty. The <u>endogenous</u> school, led by cytologists, regard cancer as a state of the cell itself and tend to become preoccupied with the desire for normalcy of cell karyology in all the microscopically observable minutiae. The significance of visible alterations in the karyosome, though commonly seen in increased extent in cancer cells, is probably more mythical than real since some frank cancer cells may display normal cell diploidy and some normal cells, even in primary cell culture, may display marked aberrations (see 1-3, 34-36). The value that derives from these two schools is that they both focus on an aberration in cellular genetics and, in turn, on its workhorse - deoxyribonucleic acid or DNA, whether visible or invisible. <u>The dragon</u> in the cell culture system then <u>is DNA</u> and the enemy has been <u>identified</u>! No one really likes dragons and it should therefore be fair game to set about destroying them - both exogenous and endogenous dragons and no matter where they hail from.

Now, dragon killing is a specialized art that lies within the domain of the biochemist since he alone has all the specialized tools and paraphernalia in his toolshed for dragon detection and destruction and he stands ready to test his mettle. Perhaps the most obvious approach to dragon killing is to find the dragon, chop it up into bits and pieces, filter out the pieces that might still be capable of wiggling, and then to apply some fail-safe system such as treating the remains with noxious chemicals and embalming fluids so that they remain dead forever. They must not, like Phoenix, arise from the ashes to live again. The analogy is that one should eliminate all possible functional DNA from the biologic product.

The approaches to biochemical dragon destruction are several. The first approach is to reduce the population of dragons one has to cope with to the smallest possible number in the first place. This may be accomplished by working with the liberated products of the cells alone with care to avoid disrupting the cells themselves. This may not be possible in those situations in which the products of the cell membranes are to be harvested and separated.

The second approach is to chop the DNA into bits by hydrolyzing it with DNA'se that has been shown to reduce the mean size to 4 nucleotides long (37,38). Dissociation from histones prior to enzyme treatment can be achieved, if necessary, by appropriate alteration in pH and ionic strength(39). Just what minimal length of DNA is required to constitute a dragon is unknown but even eliminating the need for initiators and terminators, the nucleotide chain must surely require at least a hundred or more nucleotides in properly ordered sequence in order to code for anything that is meaningful (40). At least 3 nucleotides would be required for each of the two complementary strands to provide the brush ends needed for its incorporation into the cell chromosomes. The probability for chance bringing together of the required polymerases, the required length of DNA after DNA'se treatment, the required endonucleases, and the chance fit of the brush ends into a chromosome gap seem so highly improbable an event as to render the problem a non-problem from the point of human safety. It is of importance in this regard that while nucleic acids from oncogenic primate herpesvirus (41), from SV_{40} virus (42), and from bovine adenovirus type 3 (43) have been found to possess the same oncogenic capability as the whole virus, this activity is readily destroyed by treating the nucleic acid with DNA'se (42,43), at least for SV_{40} virus and adenovirus in which tests have been made.

The third approach to assuring safety from any viable dragon's teeth or other dragon scraps that might theoretically remain in the product after DNA'se treatment is by filtration through molecular sieves or by affinity chromatography using substances whose gustatory apparatus is specially attuned to immobilization of DNA.

The fourth application in a fail-safe system is to treat the remains with formaldehyde. Formaldehyde inactivates viruses by reacting both with cell proteins and with nucleic acids (44). In the latter instance, formaldehyde reacts with the amino and imino groups of the purines and pyrimidines precluding their hydrogen bonding with the oxy and imino groups on the complementary base (45). Formaldehyde cannot react with these groups in complementary linked strands (46) but can do so in any single stranded DNA that might be present. As noted above, treatment with DNA'se reduces the DNA to such small segments as to preclude their presenting the possibility for meaningful readouts and to render them incapable of retaining double strandedness.

Finally, meaningful tests for safety prior to application to man are available. These include (a) sensitive tests to detect resident DNA, especially when tested in high concentrations, (b) the application of in vitro tests for transformation as e.g., the human B lymphocyte transformation test (47,48), and (c) tests in animals for oncogenicity.

Diligent application of the approaches now available should render any line cell capable of being established as safe. Here, indeed, is where the emphasis must be placed and pursued during the coming decade.

RESOLVE

If one looks back at cells and biologics in their historic perspective, one may wonder how primary cells became the acceptable cell to the exclusion of line cells and cell strains, especially in view of all the indigenous virus problems that some of the primary cells have generated. The only reasonable answer that can be given must be the relative state of ignorance of the art and the degree of political clout each of the decision makers could muster at the time in favor of his own theory or prejudice. It is interesting to speculate that had the relative time sequences for events been altered, we might be faced just now with any equally senseless policy of permitting the use only of line cells to the exclusion of primary cell cultures and cell strains. The big message, however, is that a new era of technology is upon us that will demand the application of line cells for their practical solution. The needs of the people and the economics of the situation demand discard of the cherished dogma of the past and the substitution of a new era in biologics production during the next decade as the supporting data base is evolved. To paraphrase a conventional wisdom, "the proof of the dragon is in the eating."

ACKNOWLEDGEMENT

The author is indebted to Dr. Alfred A. Tytell for his advice and counsel in the art of biochemical dragon killing.

REFERENCES

1. Hilleman, M.R. Cells, vaccines, and the pursuit of precedent. National Cancer Institute Monograph No. 29, pp 463-469. 1968.
2. Hilleman, M.R. Cells, vaccines, and safety for man. Biohazards in Biological Research. Edited by A. Hellman, M.N. Oxman, R. Pollack. Cold Spring Harbor Laboratory, 1973, pp 81-95.
3. Hilleman, M.R. Prospects for vaccines against cancer. Viruses, Evolution and Cancer. Edited by E. Kurstak and K. Maramorosch. New York, Academic Press, Inc., 1974, pp 549-560.
4. Hilleman, M.R. and Werner J.H. Recovery of new agent from patients with acute respiratory illness. Proc. Soc. Exp. Biol. Med. 85:183-188, 1954.
5. Biological Products, Public Health Service Regulations, Title 42, Part 73. U.S. Department of Health, Education, and Welfare, Publication 437 (revised Jan. 1967).
6. Hayflick, L. and Moorhead, P.S. The serial cultivation of human diploid cell strains. Exp. Cell Res. 25:585-621, 1961.
7. Continuously cultured tissue cells and viral vaccines. Report of a committee on tissue culture viruses and vaccines. Science 139:15-20, 1963.
8. Hayflick, L., Moorhead, P.S., Pomerat C.M. et al: Choice of a cell system for vaccine production. Science 140:766-768, 1963.
9. Sabin, A.B. Cancer and human cell vaccines. Science J A5:33, 1969.
10. Hayflick, L. Dangerous vaccines. Science J A5:33, 1969.
11. Sabin, A.B. Gamma globulin prophylaxis; inactivated rubella virus; production and biologics control of live attenuated rubella virus vaccines. Discussion on Session V. Amer. J. Dis. Child 118:378-381, 1969.
12. Perkins, F.T. and Hayflick, L. Cell cultures. Science 155: 723-724, 1967.
13. Andreopoulos, S. Human vaccines and bureaucracy: Stanford professor's odyssey. Stanford, MD 11:18-22, 1972.
14. Petricciani, J.C. Vaccine developments and the public interest. Stanford, MD 11:13-16, 1972.
15. Petricciani, J.C., Milstien, J.B., Seifried, A.S., et al. Safety of viral vaccine cell substrates: A reevaluation. J. Natl. Cancer Inst. 57:915-919, 1976.
16. Petricciani, J.C., Hopps, H.E., and Lorenz, D.E. Human virus vaccines: Why monkey cells? Science 176:813-14, 1972.

17. Minutes of the Third Meeting of the Committee on Cell Cultures, International Association of Microbiological Societies, Permanent Section of Microbiological Standardization, Philadelphia May 18, 1966.
18. Developments in Biological Standardization. Volume 37. Edited by F.T. Perkins and R.H. Regamey. Switzerland, S. Karger, 1977.
19. Girardi A.J., Sweet, B.H., Slotnick, V.B. et al. Development of tumors in hamsters inoculated in the neonatal period with vacuolating virus, SV_{40}. Proc. Soc. Exp. Biol. Med. 109: 649-660, 1962.
20. Gear J.S.S., Cassel G.A., Gear A.J., et al. Outbreak of Marburg virus disease in Johannesburg. British Med. J. 4: 489-493, 1975.
21. Cockburn, W.C. Poliomyelitis vaccination in tropical countries. Advances in Experimental Medicine and Biology. Edited by A. Kohn and M.A. Klingberg. New York, Plenum Press, 1972, pp 223-236.
22. Hilleman, M.R., Bertland, A.U., Buynak, E.B., et al. Clinical and laboratory studies of hepatitis B surface antigen vaccine. Viral Hepatitis. Edited by G.N. Vyas, S.N. Cohen, and R. Schmid. Philadelphia, Franklin Institute Press, 1978.
23. MacNab, G.M, Alexander, J.J., Lecatsas, G., et al. Hepatitis B surface antigen produced by a human hepatoma cell line. Br. J. Cancer 34:509-515, 1976.
24. Memorandum: Interferon and other antiviral agents, with special reference to influenza. Bull WHO 56:229-240, 1978.
25. Hirsch, M.S. Interferon - its hour come at last? N. Eng. J. Med. 298:1022-1023, 1978.
26. Merigan, T.C., Rand, K.H., Pollard, R.B., et al. Human leukocyte interferon for the treatment of herpes zoster in patients with cancer. N. Eng. J. Med. 298:981-987, 1978.
27. Strander, H. Mogensen, K.E., and Cantell, K. Production of human lymphoblastoid interferon. J. Clin. Microbiology 1: 116-117, 1975.
28. Zoon, K.C., Buckler, C.E., Bridgen, P.J., et al. Production of human lymphoblastoid interferon by Namalva cells. J. Clin. Microbiology 7:44-51, 1978.
29. Andersson-Anvret, M. and Lindahl, T. Integrated viral DNA sequences in Epstein-Barr virus-converted human lymphoma lines. J. Virology 25:710-718, 1978.
30. Pagano, J.S. The Epstein-Barr viral genome and its interactions with human lymphoblastoid cells and chromosomes. Viruses, Evolution and Cancer. Edited by E. Kurstak and K. Maramorosch. New York, Academic Press, Inc., 1974, pp 79-116.
31. Pritchett, R., Pedersen, M., and Kieff, E. Complexity of EBV homologous DNA in continuous lymphoblastoid cell lines. Virology 74:227-231, 1976.

32. Doerfler, W. Integration of the deoxyribonucleic acid of adenovirus type 12 into the deoxyribonucleic acid of baby hamster kidney cells. J. Virology 6:652-666, 1970.

33. Botchan, M., Topp, W., and Sambrook, J. The arrangement of Simian virus 40 sequences in the DNA of transformed cells. Cell 9:269-287, 1976.

34. Harnden, D.G. Chromosome abnormalities and predispostion towards cancer. Proc. Roy. Soc. Med. 69:41-43, 1976.

35. Levan A., Levan, G., and Mitelman, F. Chromosomes and cancer. Hereditas 86:15-30, 1977.

36. Chromosomes and cancer. The Lancet ii:227-228, 1977.

37. Deoxyribonuclease I (Bovine pancreas). Worthington Enzyme Manual. Edited by L.A. Decker. New Jersey, Worthington Biochemical Corporation, 1977, pp 153-155.

38. Laskowski, M., Sr. Deoxyribonuclease I. The Enzymes. Third Edition. Edited by P.D. Boyer. New York, Academic Press, Inc., 1971, pp 289-311.

39. Smart, J.E. and Bonner, J. Studies on the role of histones in relation to the template activity and precipitability of chromatin at physiological ionic strengths. J. Mol. Biol 58:675-684. 1971.

40. Harper H.A., Rodwell, V.W., Mayes, P.A., et al. Protein synthesis and the genetic code. Review of Physiological Chemistry. Sixteenth Edition. California, Lange Medical Publications, 1977, pp 433-447.

41. Fleckenstein, B., Daniel, M.D., Hunt, R.D. et al. Tumour induction with DNA of oncogenic primate herpesviruses. Nature 274:57-59, 1978.

42. Sol, C.J.A. and van der Noordaa, J. Oncogenicity of SV_{40} DNA in the Syrian hamster. J. Gen. Virol. 37:635-638. 1977.

43. Igarashi, K., Sasada, R., Kurokawa, T., et al. Biochemical studies on bovine adenovirus type 3. IV. Transformation by viral DNA and DNA fragments. J. Virol. 28:219-226. 1978.

44. Staehelin, M. Reaction of tobacco mosaic virus nucleic acid with formaldehyde. Biochimica et Biophysica Acta 29: 410-417, 1958.

45. Feldman, M.Y. Reactions of nucleic acids and nucleoproteins with formaldehyde. Progress in Nucleic Acid Research and Molecular Biology. Volume 13. Edited by J.N. Davidson and W.E. Cohn. New York, Academic Press, 1973. pp 1-49.

46. Sarkar, N.K. and Dounce, A.L. A spectroscopic study of the reaction of formaldehyde with deoxyribonucleic and ribonucleic acid. Biochimica et Biophysica Acta 49:160-169, 1961.

47. Pope, J.H., Horne, M.K., and Scott, W. Transformation of foetal human leukocytes in vitro by filtrates of a human leukaemic cell line containing herpes-like virus. Int. J. Cancer 3:857-866, 1968.

48. Katsuki, T. and Hinuma, Y. Characteristics of cell lines derived from human leukocytes transformed by different strains of Epstein-Barr virus. Int. J. Cancer 15:203-210, 1975.

VIRAL VACCINE SUBSTRATES

Introductory Remarks

John P. Fox, M.D.

Department of Epidemiology
University of Washington
Seattle, WA 98195

By title, this second session is concerned with substrates for viral vaccines and, more specifically, with cell culture substrates rather than embryonated eggs or CNS tissue from intact hosts. However, except for Dr. Netter who provided a review of recommendations and WHO requirements for producing inactivated viral vaccines in diploid cell lines, the speakers extended the scope of their presentations substantially beyond that defined by the session title.

Dr. Osborn's presentation dealing with vaccines under development is focused much more on the principles and process of vaccine development than on an overview of all developing vaccines. Of the latter, a selected but important few are used for purposes of illustration. She opened with a provocative presentation of significant advances in relevant basic sciences and of their implications for vaccine development. In this context, cell substrates are mentioned chiefly in relation to viral pathogens for which nonavailability of a suitably permissive cell substrate is a major factor limiting progress. One potentially important mechanism in developing new live virus vaccines is referred to as "host-induced modification of virulence" and illustrated with rather elegant modern examples. Although the mechanisms still are not well understood, there are a number of earlier examples of virulence modification resulting from adapting viruses to unnatural hosts. These include yellow fever and measles.

Dr. Osborn also makes a forceful argument <u>for</u> rational design and <u>against</u> empiricism in both developing new vaccines and improving existing successful vaccines, the first generation of which, in the eyes of Dr. Harry Meyer, was developed almost purely empirically.

Actually, the development of one of these, yellow fever vaccine, could be cited to reinforce her argument since, in the light of then existing knowledge, its development was highly rational (and, belatedly, resulted in a Nobel Prize for Max Theiler). Briefly, on serial intracerebral passage in mice, yellow fever virus loses its viscerotropism (ability to attack the liver) but gains in neurotropism. Theiler reasoned that prolonged propagation of mouse-adapted virus in non-nerve tissue (cultures of minced, headless chick embryos) should yield a strain devoid of both virulence tropisms as tested in rhesus monkeys.

The presentation of Dr. Beale directly addressed the central problem of the Symposium, namely the choice of cell substrate for biological products (not just viral vaccines) for use in man. The critical point is the yield of the desired product (hormone or, perhaps, antibody in the future; the interferon or virus in the present). For both of the latter, readily available heteroploid cell lines produce very high yields, are easily adapted to large scale production methods and, hence, promise great economy in production. On all these counts such cell lines are far superior to other possible substrates such as acceptable primary cells or diploid cell lines. Reluctance to accept the products of such substrates derives from their potential contamination with a transmissable agent associated with or responsible for the transformed state of the substrate cells and which might be oncogenic in man. Since direct testing for oncogenicity in humans is neither ethical nor technically feasible (a very long but indefinite period of observation would be needed), a different form of evidence is required to resolve the reluctance.

Dr. Beale has made a strong case for overcoming this reluctance using interferon and poliovirus for an inactivated vaccine as examples. This case is based in part on precedent provided by extensive past use, with no recognized oncogenic effects, of both live (yellow fever, polio) and inactivated (influenza, polio) vaccines produced with minimal purification in substrates now known to have been contaminated with oncogenic viruses. In perhaps more compelling part, the case is based on results of testing the products by highly sensitive procedures for undesired contaminants such as nucleic acid and infective viruses.

In the final presentation, Dr. Salk recalled previous discussions on the theoretical risks associated with the use of tumorigenic cells in the production of vaccines. He went on to suggest that there are experimental approaches that one might take to develop data to answer the question of theoretical risks. Dr. Salk also pointed out that it might be possible to generate specific data on human risk if follow-up studies were undertaken on human subjects who had received vaccines produced in tumorigenic cells more than 20 years ago.

VIRAL VACCINES UNDER DEVELOPMENT: A THIRD GENERATION

June E. Osborn, M.D.

Department of Medical Microbiology and Pediatrics
497 Medical Sciences Building
University of Wisconsin Medical School
Madison, Wisconsin 53706

INTRODUCTORY COMMENTS: THE NEW GENERATION

These are exciting times in virology, immunology and molecular biology, and there is a brisk pace of advance in the basic sciences which provide tools for vaccine development. Additionally, the brilliant success of vaccines created in the fifties and sixties has spawned a host of attempts to apply active immunization principles to the prevention of numerous viral diseases which still evade efforts at control. Partial reviews of current vaccine candidates have appeared in profusion in recent months, and so I have modified my initial purpose for this presentation. Rather than attempting a compendious "state of the art" effort concerning the candidate vaccines themselves, my aim here will be twofold. First, I wish to highlight some of the recent scientific advances which can and/or should be elements in rational vaccine development. Second, I will assess several of the current candidate vaccines in the context of new knowledge with the goal of identifying factors that constitute limits to their further improvement. I should in good conscience confess at the outset that my attention to cell substrates *per se* will be somewhat peripheral except as I perceive them to be the central factor which limits progress in a given context.

Dr. Harry Meyer has often talked of the presently available viral vaccines in terms of generations(1). In his parlance, the first generation of vaccines were developed almost purely empirically in the pre-tissue culture era.

In that generation, therefore, the limitations of empiricism were exaggerated by the lack of suitable host cells for production and testing of vaccine candidates. The second generation, in Dr.

Meyer's construction, resulted from the explosion of enthusiasm which accompanied the recognition of the power of tissue culture technology in virus research; assaults were mounted against most of the highly visible, acute viral diseases which had not been attacked in the first round. In this second effort the dependence on tissue culture technology meant that cytocidal viruses were the earliest to yield; even in the case of rubella vaccine, the absence of cytopathology induced by another virus became the end point for identifying productive rubella infection.

The impressive progress achieved during the first two generations of viral vaccine development resulted in the proud fact that effective prevention through active immunization could be achieved against smallpox, yellow fever, poliomyelitis, measles, mumps and rubella. While influenza vaccine was somewhat less satisfactory, advances in biochemical technology and immunology facilitated distinct improvements in more modern versions of the influenza vaccine.

Nevertheless, while this progress was on-going and deserving of great pride, new viral pathogens were being identified in bewildering profusion and knowledge of host-virus interactions achieved levels of complexity which dified simple serologic assessments of vaccine efficacy. It is interesting, therefore to take a look at the new crop of vaccine candidates from as many vantage points as possible: in many instances, empiricism is no longer the only approach that can or should be taken to their development (2,3,4). Furthermore, their importance as etiologic agents of disease is in some cases obscure or moot. Increased stringency in human subjects usage - particularly in the context of nontherapeutic research on children (5-8) - commands that issues of safety and need be clearly addressed and that animal model experimentation be maximized in the process of vaccine development. Finally, scarcity of some research animals and concern about the use of others adds a further stricture on progress (9,10) in the creation of new viral vaccines for mankind.

RECENT BASIC SCIENCE PROGRESS PERTINENT TO VIRAL VACCINES

Before turning to specific vaccines under development, let me illustrate some of the basic science advances I am referring to which might soon submit to rational application for purposes of vaccine design. Starting with cell science, a number of provocative observations suggest that the future may allow much more manipulation of virus-host relationships than is presently possible in cell cultures.

CELLULAR GENETICS

For instance, one such area of progress is the role of cellular genetic control in permissiveness to viral infection. Initial use of somatic cell hybrids and cell fusion techniques gave strong suggestion that genetic variation could markedly affect cellular resistance to viruses. Recently the control of interferon production has been assigned some quite specific genetic loci on the human chromosomes 2,5 and 21(11), and multisomy of chromosome 21 has resulted in predictable amplification of the interferon response. Since our interest is in vaccine development, and presumably, increased host cell permissiveness, that finding might seem to be in the wrong direction, but I perceive it as an early product of a burgeoning genetic exercise: that of mapping the human (and other mammalian) chromosomes for gene function.

Host Gene Products Which Influence Permissiveness

Eukaryotic genes have begun to assume increasing relevance in regulation of viral productivity. Until a few years ago, the question of cellular permissiveness seemed to depend entirely on the presence or absence of specific receptors on cellular membranes. The specificity of these receptors was impressive: for instance, among Coxsackie B viruses, the odd-numbered ones interfered with themselves but not with the even-numbered ones(12). A dominant role for receptors as a main determinant of permissiveness was further suggested by the relatively early demonstration that poliovirus RNA could successfully transfect non-primate cells once the capsid proteins were removed. There were hints from rather complex dual-infection experiments that observed interference between viruses could be mediated by additional cellular controls, but these were usually interpreted to be shortages of shared substrates or sites on ribosomes rather than specific gene product functions.

Subsequently, a greater complexity of cellular control was revealed in circumstances where viral initiation of multiplication was intimately keyed to the cell growth cycle. For instance, papovaviruses replicate in concert with cellular DNA synthesis (13) and human cytomegalovirus also replicates in parallel with cellular DNA (14) although in the latter instance there is some argument as to whether there is any actual cytomegalovirus dependence on cellular DNA synthesis (15). It is clear, however, that the physiologic state of cells plays a substantial role in human cytomegalovirus efficiency of replication (16). In the case of retroviruses, of course, their productive replication is entirely dependent on cellular DNA replication (17).

A more obscure finding but one suggestive of additional cellular controls was the observation that pretreatment of cells with iododeoxyuridine or other DNA synthesis inhibitors actually augmented the subsequent replication of cytomegalovirus (18), inferring that unidentified cellular inhibitors were not synthesized under these conditions allowing more efficient viral productivity. This kind of observation is quite hopeful for cell science in the service of virology, for it allows one to postulate the existence of presently unknown gene products which may submit to precise manipulation in the future to improve host cell permissiveness.

Host Induced Modification of Viral Virulence

The kinds of phenomena I have been describing are all explicable within the bounds of sharply separated cellular and viral genetic functions. However, recently a quite new concept of host role in animal virus replication has emerged, which can be expressed in the phrase "host-induced modification of virulence." Clearly in bacteriophage systems the bacterial host can play a significant role in the actual biochemical structure of the progeny phages by means such as specific base methylation which in turn affects virulence of the progeny for subsequent hosts (19,20).

In animal virology only one rather artificial example of such host interplay was extant in the literature until recently; Ishida and Homma had demonstrated in 1960 that Sendai virus produced in eggs was avirulent for L cells whereas that produced in L cells was virulent for both host systems (21). Their report drew little attention, nor was there much extrapolation from bacterial models until 1974 when Scheid and Choppin demonstrated that post-translational cleavage of envelope glycoproteins of paramyxoviruses was a necessary step before progeny virions could express full virulence in subseqent cells (22). They further demonstrated that some host cells could provide enzymes for the specifically required enzymatic cleavage and therefore produced fully virulent virus, while other cells lacked the appropriate enzymes for cleavage and therefore produced virions which were avirulent unless their glycoproteins were processed *in vitro* by addition of enzymes (23).

These demonstrations of host-induced modification of virulence are of potentially great importance, for they broaden the range of possible virus-host interactions and could spark a series of investigations concerning host role in viral maturation and virulence. Again, I think the implications for advances in cell science as it relates to virology are evident.

In our laboratory an animal model system has yielded findings which suggest that host-induced modification of virulence may occur in the intact animal host and may be a critical factor in considering live viral vaccine candidates for chronic, persistent viral infections (24). Murine cytomegalovirus in mice establishes a persistent, productive infection of submaxillary glands (25); and murine CMV prepared by extracting virus from suspensions of salivary gland tissue is our accustomed source of virulent virus in the animal model system we have employed for the study of acute CMV infection. Among its virulence properties are the ability to multiply in reticuloendothelial organs such as spleen (26), to induce transient immunosuppression during acute infection (27,28), and to kill newborn mice rapidly (26). All of these virulence properties are ablated by one (or more) passages of MCMV through mouse embryo cell cultures; the resultant "avirulent" virus no longer multiplies in spleen, does not cause immunosuppression, and seems not to affect newborn mice until two weeks or more after inoculation. At that time, however, productive salivary gland infection has again been established and the animals begin to die belatedly with high titers of MCMV in their spleens (24; R.R. Ryshke and J.E. Osborn, unpublished observations). The best-fit hypothesis to explain this peculiar set of findings is that some facet of submaxillary gland cell processing or packaging of MCMV is essential for full expression of viral virulence - and thus there is a suggestion in this model that the importance of host-induced modification of virulence may not be restricted to cell culture phenomena but may also be a consideration in refining our appreciation of virus-host interactions in the intact animal.

Genetics of Animal Viruses

Thus far we have been considering host cell genetic properties and products as they pertain to permissiveness and viral virulence. The field of animal virus genetics per se has been something of a swamp until recently, but the combined impact of restriction enzyme and nucleic acid sequencing technologies along with temperature sensitive mutant manipulation and complementation grouping has opened the way for precipitous escalation of knowledge of the genetics of viruses. For instance, the entire genetic sequence of the SV 40 genome was published recently (29) and the recently discovered BK human papovavirus is followed quickly. Even the huge and complex genome of herpes simplex has yielded much precise information through the patient sorting of mutants by Priscilla Schaffer's group (30) and the elegant restriction enzyme mapping and polypeptide studies of Bernard Roizman's laboratory (31a). The latter efforts have yielded results of sufficient refinement that it has been possible to do "molecular epidemiology" studies, using the highly individualized restriction fragment "finger prints" and polypeptide patterns to track specific herpes simplex isolates through a human population (31b). The potential usefulness of this kind of technique is evident when you realize that it adds an even more refined means of viral recognition than did antibody neutralization kinetics, and allows one to sort antigenically identical isolates into distinctive subsets.

A related example of the improved precision afforded by advances in viral genetics and molecular biology can be illustrated by recent work with polioviruses. Changes in the VP1 protein gene product have been shown to correlate very closely with attentuation of all three serotypes (31c; O.M. Kew and R.R. Rueckert, personal communication), and prospects seem bright for asking refined questions about successive attenuation and rate of reversion to virulence in various host cells by precise tracking of VP1 and, possibly, other markers yet to be found. Thus far the observations of variation in VP1 are simply correlative but nonetheless attractive, for if specific biochemical markers of attenuation were to become available, the unwieldy neurovirulence testing currently in place might yield to a much more reproducible, rational and streamlined test for assuring safety of attenuated poliovirus strains.

Biomedical Ethics

Turning from cell genetics and molecular virology, in which fields advances have generally held promise of simplifying life for the vaccine developer, the maturation of two other fields has made life much more complicated. The fields to which I refer are biomedical ethics and immunology. Addressing ethics first, it is fair to say that things seemed simpler ten years ago or more than they do now. The first and second-generation vaccines were - at least at the time of their development - approached with conceptual ease; in general they were directed against what were perceived to be acute, self-limited viral infections of unarguable importance. Their efficacy could be monitored by the appearance of circulating antibody which clearly signalled resistance to secondary attack. Thus the benefit/risk ratio was sufficiently favorable that the use of human subjects in vaccine testing and eventual licensure seemed to pose relatively little in the way of ethical problems. Unquestionably, biomedical ethics was at a substantially less sophisticated stage of conceptual development than it is now.

The change of the times was signalled by the increased clamor attending the use of rubella vaccine which, at least in the United States, was directed primarily against persons in whom the disease would have been mild, in order to protect persons other than the vaccine recipients. The active immunization of someone for the protection of someone else is not at all a foreign concept to public health professionals, but it certainly agitates the ethicists! I point that out not to start the argument again but because the issues that initially surfaced with rubella intensify greatly in the context of some of the new vaccines I want to consider. To the extent that diseases are less virulent or vaccines are less predictably effective in prospect, ethical issues of the use of human subjects - especially children - in nontherapeutic research become troublesome in this modern era.

These issues are not readily resolved, but one possible consequence of the increased complexity surrounding human subjects usage for research has been an intensified awareness of the potential usefulness of animal model systems. This pressure to glean all one can from animal models is welcome, for much manipulative experimentation is made possible when suitable animal models exist. However the hazard must always be kept in view that inaccurate extrapolation between species can be grossly misleading. A mouse is demonstrably not a man, and for that matter neither is a chimpanzee.

Viral Immunology

When it comes to making life tough for vaccine developers, however, the ethicists can't hold a candle to another group: for immunologists have really been muddying the waters lately, and the viral immunologists have done much of the splashing. The conceptual complexity of pathogenesis and protective immunity introduced by recent advances in immunology is sufficiently awesome that I hesitate to approach it even in passing. Just listing the facets of immune response that have been delineated in the past decade or so is dizzying: secretory immunoglobulins seemed relatively easy to accept, but they were quickly followed by T cells, B cells, killer cells, null cells, T- and B-cell interactions, T cell and macrophage interactions, transfer factor, complement dependent or independent antibody mediated cellular cytotoxicity, *de novo* interferon responses, immune interferon responses, direct complement-mediated viral inactivation, antigen capping and antigenic modulation, and the whole issue of the role of histocompatibility antigens in susceptibility to viral infection. Perhaps the unkindest cut of all was the discovery of *infectious* immune complexes in herpes simplex and other systems, for even old, familiar neutralizing immunoglobulin G can be circumvented in such circumstances (32). Finally, while infectious nucleic acids are still a laboratory trick, the awareness that naked DNA of herpes simplex (33) or cytomegalovirus (34; A.F. Lakeman and J. E. Osborn, unpublished observations) can in fact "transfect" a cell proves a potential route for a further end-run of our accustomed immunologic barriers.

In any event, all of these discrete immunologic phenomena may play roles of varying importance in host response to a given pathogen. It is not surprising that the elements of protective immune response seem to be especially complex in chronic and persistent viral infections against which many of the candidate vaccines are aimed. My main point is that the days are fast passing in which a simple measurement of IgM and IgG specific antibody responses to viral antigens will suffice to assure protection against a pathogen; and the question of which of the profuse components of immune response are pertinent to actual protective efficacy should be addressed to both old and new vaccines alike.

A single provocative example may suffice: it has been reported in the context of both herpes simplex and varicella zoster that the immunologic response is differentially directed against envelope or capsid antigens in primary versus secondary infection. Since cell-to-cell spread may well explain reactivation in the face of specific immunity, one might conjecture that primary infection would invoke greater response to envelope antigens and secondary response to capsids. That is satisfactorily the case for herpes simplex but precisely the opposite is true for varicella-zoster (35). In each case the difference between primary and secondary responses is sufficiently exclusive, in any event, that in the rational design of a vaccine for these agents a biological engineer would certainly want to assure that effective antigenic stimulation related to both sets of antigens - and it is not at all clear that a simple, empirical approach would achieve that end, except perhaps fortuitously.

AN ANALYTIC LOOK AT CURRENT VIRAL VACCINE DEVELOPMENT: HOW FAR HAVE WE MOVED FROM EMPIRICISM?

This lengthy introduction has been intended to lay a groundwork for an analytic perspective on a few of the vaccines which content for membership in the "third generation" of viral vaccines. By now it should be evident that I have a bias against empiricism and would contend that the same kind of analysis should be applied with vigor even to the highly successful first- and second-generation vaccines.

Questions

The questions at issue are several. Why are they attenuated? Attenuated with respect to which virulence properties? Are there other facets of infection by the parent pathogen which are not addressed by the attenuated, derivative agent? (Similar questions can be slightly rephrased to pertain to inactivated or subunit vaccines; in all cases the remaining questions are germane). What are critical variables in the host response? And finally, having accurately answered all those questions, is vaccine development really worth it in terms of benefit/risk analysis or are there other more pertinent approaches to control of the disease in question?

Specific Vaccines to be Considered

Live, attenuated temperature-sensitive influenze A. Since I am extolling rationality in vaccine development, I should perhaps start with vaccine efforts which epitomize that approach; both influenza A and respiratory syncytial viruses have been submitted to a careful process of selection hinging on the use of temperature sensitive viral mutants and nasopharyngeal delivery of vaccine. The underlying theories are that local respiratory tract immunity is/should be important and that viruses for which 33°C is permissive and 37°C is not will replicate adequately in the upper respiratory tract and fare poorly as they descend into more pathogenic territory. This is a line of reasoning - to use a phrase a colleague of mine often applies to my most enthusiastic hypotheses - that is so appealing that it deserves to be right. In fact, influenza A temperature sensitive vaccines have already enjoyed widespread usage in other countries and show some promise in experimental circumstances in the United States (36-38). However, it is disappointing to learn that the efficacy of ts flu vaccines does not greatly improve on our experience with the empirically derived inactivated products used here. The balance between attenuation and complete crippling of mutant strains turns out to be delicate. What is more, it is sobering to note that, despite the awesome elegance of recent molecular studies of the segmented influenza A genome (39), the sole allowable conclusion to date regarding influenza virulence is that it is a polygenic characteristic - which is a sophisticated way of saying we still don't know what conveys virulence properties to a given influenza A strain. These present problems with live, attenuated flu vaccines should, however, be transient, for the science underlying their solution is highly developed and the rationality of this approach remains most attractive. I will go no further in a discussion of flu vaccines, since recent, capable reviews are available (38,40).

Respiratory syncytial virus vaccines. The efforts to develop a respiratory syncytial virus vaccine have also been reviewed recently (41) but they are sufficiently interesting to warrant quick treatment here. The initial experience with a killed virus vaccine was a sobering one (42) which has been a useful stimulus for caution in approaching viral diseases with a complex immunologic base. The respiratory syncytial vaccine experience was not unique, and the general topic of disease accentuation after immunization with inactivated microbial vaccines was dealt with at a workshop at the National Institutes of Health which has been tersely and adeptly reviewed by John Craighead in the Journal of Infectious Diseases in 1975 (43).

With the unhappy experience of inactivated respiratory syncytial virus vaccine in mind and the awareness that passively acquired maternal antibody might even be a deleterious element in infant respiratory syncytial virus infection, further efforts at vaccine production were attempted with caution and from a different direction (44). That a vaccine was needed became increasingly clear as the extent of respiratory syncytial virus pathogenicity in infants was recognized (45-47). Studies have therefore been directed at creation of live virus vaccines against r.s. virus.

Early work by Chanock's group first resulted in the development of two candidate temperature sensitive mutant vaccines which failed to plaque at 37°C, which were administered in early trials by the nasopharyngeal route. One of these seemed overly attenuated and the second while quite immunogenic was deemed unacceptably "hot" since it produced bothersome rhinorrhea in most recipients. That group has subsequently mutagenized the latter strain further and are working with two clones whose ability to plaque is lost at an even lower temperature than their parent.

Meanwhile, Hilleman's group have a candidate attenuated respiratory syncytial virus strain which they have chosen to apply parenterally; their initial tests have shown promising immunogenicity. Thus the field of r.s. vaccine development is lively. It is important to note, however, that despite the honest efforts of the vaccine developers to grapple with problems of r.s. vaccine usage, the fact that the disease and positive immunologic response are linked in the young infant age group for which the vaccine is intended continues to present a major conundrum in analyzing its ultimate usefulness.

Parenthetically, respiratory syncytial virus represents a circumstance in which cell substrate limitations have contributed substantially to the relatively slow rate of progress; for extracellular yields of r.s. virus are low and much basic molecular virology remains unexplored. While the vaccine efforts are yielding promising results, it seems evident that a truly permissive host cell system which allows high yields of purified virions would facilitate much refinement and conceivably better definition of antigens pertinent to the protective (and damaging) facets of the immune response.

HEPATITIS B "SUBUNIT" VACCINES

Hepatitis B is another disease for which a useful vaccine would be highly desirable. In this instance, both cell substrate and animal model considerations represent serious drawbacks to rapid, rational progress. Since good recent reviews of the status of hepatitis B have been written by persons much better qualified than I am (49-53), I will limit my comments to a few observations.

First, the candidate hepatitis B vaccine is, in essence, a prototype of another new approach to active immunization - *viz.* subunit vaccines. In the instance of hepatitis B, the subunit in question is the 22 nm particle purified from serum and freed of potentially contaminating infectivity by two or more inactivation steps. This suboptimal source is of course obligatory since no permissive *in vitro* cell system has yet been identified. The rationale behind subunit vaccines in general is that the full viral genome may convey oncogenic or other unwanted information as well as potentially replicating to produce disease in the inoculated host; and in the case of hepatitis B this concern is raised by observations of greatly increased frequency of primary hepatomas in countries where endemic hepatitis is common(54).

A second facet of hepatitis B study worthy of comment is the status of the animal model system. While chimpanzees can be infected with a demonstrable response to antigen, they do not reproduce the severe, life threatening disease pattern which one wishes most to prevent in man; and so protective efficacy must be extrapolated from serologic responses (55,56).

Third, the immunologic response to hepatitis B during natural infection is incompletely understood. Detailed analysis of serologic responses has helped to dissect at least two serotypes, but the lack of complete virions in abundance has sharply limited a comprehensive analysis of cellular immune responses. Since there is a possibility that cellular immunity can in fact contribute to the inflammatory liver disease itself, this lack of understanding gives one pause and narrows the usefulness of the candidate vaccine to patently high risk groups in which the benefit/risk analysis is intuitively favorable.

HERPESVIRUS VACCINES

Varicella-zoster

After those forays into less familiar territory, I would like to focus attention on terrain to which I am more accustomed - for much developmental activity is occurring in the realm of herpesvirus vaccines. Foremost among these from several points of view is varicella-zoster virus vaccine (57).

When I first learned of the efforts of Dr. Takahashi and his group at Osaka to develop a live, attenuated varicella-zoster vaccine I confess to a sense of both surprise and alarm. The surprise derived from my perception that production of cell-free V-Z virus in any reasonable quantity remained an impossible technical feat except in highly impractical cell culture systems such as primary or passaged human or primate thyroid cell cultures (58,59). Efforts to improve cell-free yield had been only marginally successful (60), and in fact I had come to the conclusion that little progress would be made with V-Z virus until someone learned what maturational step was deficient *in vitro* which resulted in such strict cell-association.

My sense of alarm stemmed from the hypothesis that viral properties governing virulence in primary infection were by no means automatically linked to those required for latency and reactivation in the form of herpes zoster. In fact, it is still an unresolved issue whether the 8% of persons who develop zoster over a lifetime are different *because* they establish latent infection to begin with, or whether *all* V-Z infection results in latency and only 8% reactivate (61). In any event zoster represents a significant health problem: Sabin extrapolates that there may be 700,000 cases per year in the United States (62) and all discussants acknowledge that manipulation of the varicella-zoster virion might inadvertently enhance those properties resulting in easy reactivation. It is clear that the V-Z virus sometimes establishes stable residence in dorsal root ganglia (63) but the enthusiasts hypothesize that empirical attenuation will reduce the likelihood of that colonization. I must comment that that hypothesis is based on hope, not data.

A final source of my concern is that, with chickenpox being a rather benign if annoying disease for most children, the benefit/risk assessment seems far different, for instance, from that pertaining to a uniformly serious disease such as measles. In fact, one needs to be a bit of a "vaccine hawk" to advocate uniform use of a V-Z vaccine in all children unless the preceding concerns are thoroughly resolved.

With all that worry as background, I must confess I am beginning to join the ranks of those who are impressed with recent findings from the Japanese group (57, 63, 64). Their approach to development of the vaccine was empirical (65) - they seem to have approached the issue of attenuation by simple, repetitive passage in cell cultures of guinea pig embryo and WI-38 cells. They do not comment on the problem of cell association, but their yields of infectious virus after sonic disruption of cells are between 10^2 and $10^{3.5}$ pfu/ml, and I read between the lines that this limitation remains a major obstacle to mass production of batches of uniform vaccine.

Nonetheless they have prepared and tested their candidate vaccine in at least 180 children with impressive efficacy and no mishaps thus far. They have proceeded in cautious, step-wise fashion. Initial success at raising a serologic immune response in normal children (66) was followed by the demonstration of protective efficacy in that test group (67-69), and they have recently proceeded to the quite awesome experiment of using the live, attenuated varicella-zoster vaccine in high risk children (70). Two years later, durable immunity is demonstrable in most vaccine recipients. Altogether, their results are most impressive but leave me with some nagging concerns: keep in mind the differential antigenic response to primary and secondary V-Z infections noted earlier (35) as well as the fact that advances in interferon research suggest that alternative modes of control may soon be at hand (71). With these elements at issue, the question will remain for a long time to come: is latency and reactivation any different in this virus-host combination than with the wild virus?

In summary, however, this particular member of the third generation is maturing fast and the desire for improved control of V-Z in hospital settings and certain high risk groups is so strong that pressures for licensure are likely to be intense. It is to be hoped, therefore, that the problems attendant on its rational study will give way quickly: both the mystery of its strict cell association and the unavailability of an animal model and/or suitable range of assays of cellular immune responses will need to be solved for fully rational assessment of the vaccine, however striking its initial efficacy has been.

Cytomegalovirus

The varicella-zoster question is simplicity itself, however, compared to cytomegaloviruses. Here the elements of rational analysis are almost completely lacking as at least two groups of vaccine enthusiasts race to develop live, "attenuated" cytomegalovirus vaccines for use in man (72-74). Since acute cytomegalovirus infection is characteristically avirulent in the mature human host, "virulence" is hard to define and "attenuation" therefore is even more so (75, 76). Chronic persistent or latent cytomegalovirus infection is so clearly a common fact of life that I often refer to cytomegalovirus as part of the normal "viral flora." There is good evidence of substantial antigenic heterogeneity among cytomegalovirus isolates, and several groups have in fact tried to assign three or more serotypes to the various human cytomegaloviruses that have been studied (75, 77-80). Those serotypes, however, are not crisply defined as are, for instance, the poliovirus types, but represent sort of an antigenic "smear." Despite their close relatedness, the immunologic controls that govern their reactivation from latency are so finely tuned that there is evidence from one study that activation of one serotype may occur while others remain quiescent in the same host (81). Parenthetically, while it is a little out of the order of my sequence, there is a fascinating confirmation of this purported phenomenon with herpes simplex type 2. Nahmias and Roizman, having established the molecular restriction enzyme fragment "fingerprint" patterns I referred to earlier, have applied these to the sequential study of some patients who were being followed in a venereal disease clinic for intermittent, symptomatic herpes genitalis (B. Roizman, personal communication). The eleven patients in their longitudinal study had herpesvirus isolates that yielded characteristic, stable DNA restriction fragment patterns which could be readily distinguished from one another. During the course of that study one subject, while continuing to excrete her own characteristic herpes simplex type 2, concurrently acquired and began to excrete another HSV-2 with DNA characteristics of another patient in the study. You can draw your own conclusions about the social circumstances surrounding the transfer, but the implications for efforts at prevention are most depressing, for the superinfecting strain was antigenically identical to the endogenous virus.

But back to cytomegalovirus: all these considerations discourage me about even the theoretical potential of a useful cytomegalovirus vaccine. Even in the circumstance of congenital infection, prevention of which would surely be desirable, it must be noted that congenital infection is several times more frequent than is resultant disease (82). Mothers who deliver congenitally infected infants demonstrate a specific defect in their cellular immune response (83, 84); and sequential delivery of two congenitally infected infants has been documented on several occasions (85-88). Conversely demonstrated primary infection occurring at what might be presumed to be a critical stage of gestation has yielded a normal (although infected) offspring (89).

Furthermore, the message coming from animal model studies is also mixed. Both Howard's and Medearis' groups have demonstrated that vaccine efficacy can be established for the mouse cytomegalovirus model (90,91). In my laboratory we have also shown that in the sense that we can protect mice from severe exogenous challenge by use of our attenuated MCMV (92). However, as I described earlier, longer follow-up of those protected mice shows that their now-endogenous virus later brings them to grief as it appears to reacquire virulence while replicating in the vaccinated host. In the guinea pig cytomegalovirus model it is evident that both immune and seronegative guinea pigs can have infected offspring (93,94) although, in fairness, both guinea pigs and humans who are seropositive during pregnancy seem to have undiseased albeit infected offspring.

In my mind, all this argues for caution. But empiricism reigns and current efforts at vaccine development are based on repetitive passage in tissue culture to achieve "attenuation" (95,96). The investigators cite, as success, the ability of their candidate vaccines to raise complement-fixing antibodies in adult recipients (97,98). The vaccines have rather regularly invoked local reactions one or two weeks post-inoculation which lasted several days and would have brought howls of outrage if we were talking about influenza vaccines. Nevertheless, the developers' enthusiasm is high, their ultimate target population seems to be adolescent girls, and they have now embarked on a test of cytomegalovirus vaccine in persons two to three months before renal transplantation on the theory that they will have achieved a state of latency or innocuous persistence at the time of immunologic perturbation (99). I hope they are right.

Herpes Simplex

Probably the toughest problem of all comes with herpes simplex. Both types 1 and 2 cause sufficient grief that a substantial number of sufferers are desperate for relief. For them I see very little hope in sight, for persons who are seriously symptomatic are a small and unpredictable subset of the infected population, and they have probably made their own unsatisfactory peace with herpes simplex long since. The example I cited earlier of superinfection of a chronically diseased host makes it likely that they are beyond the reach of present active immunization alternatives. It may be that further immunologic sophistication concerning certain specific antigens which relate to refractoriness to reactivation may change that assessment. Immune interferon responses seem to be of considerable importance (100) and certainly cellular cytotoxicity and complement mechanisms as well as infectious immune complexes can be shown to be pertinent in the chronic balance between host and virus.

(As an aside - to borrow a leaf from Sidney Gellis - I'm not at all sure I can stand much more immunologic sophistication in the area of herpes simplex). Anyway, at present, immunologic manipulation of established, recurrent infection seems remote.

By contrast, a number of efforts are underway to develop vaccines effective against primary herpes simplex infection. The efforts are generally in the direction of subunit vaccines and/or other nucleic-acid free preparations (101), for the oncogenic potential of inactivated herpes genomes is well established and the live, attenuated approach raises the familiar concern about latency (102). Efforts at vaccine production for man are considerably less far along than are V-Z and CMV studies, but some reports have appeared in the literature (103).

However, intensive study of animal models is underway and an excellent review of the kinds of findings in those models has recently been written by a group from the Bureau of Biologics (102). Most of the studies involve the mouse, and there are those who protest that their relevance if any to the human infection will be serendipitous. Those persons reason that the study of a pathogen in an unnatural host is not likely to yield relevant results; there is sufficient sense in this argument to add an extra note of caution in extrapolation from mouse- and rabbit-model studies. For instance it is regularly argued from these models that less severe primary lesions are predictors of less frequent latency and reactivation - and this in fact is a main source of reassurance for the V-Z enthusiasts. Let me just point out here - as was done with great thoroughness by Parks and Rapp in 1975 (103) - that among known herpesvirus vaccines currently in use in other species, one can find examples of almost any point one wants to make. In some cases such as Marek's disease, attenuated viruses prevent disease but <u>not</u> wild virus multiplication. In other instances, more classic prevention can be shown. For vaccine enthusiasts, I will note that there are even instances among the oncogenic primate herpesviruses where the oncogenic property itself has been successfully attenuated, yielding a vaccine with protective efficacy against challenge.

With that in mind and - having thoroughly discouraged you about herpes simplex vaccines - I will conclude with a note of optimism; for it may yet be that herpes simplex types 1 and 2 will be the pathogens for which rational analysis and vaccine design prove most potent. Bernard Roizman, who has provided much of the elegant molecular biology on which our current knowledge of herpes simplex is based, is persuaded that both latency and the ability to transform cells are dependent on the function of one or a few specific gene products of the herpes genome (Roizman, personal communcation.) Were this the case, excision of those genetic loci would then allow the rational design of live, attenuated vaccines for use in the prevention of primary infection. Such an end is way off, but not necessarily a long way off.

Further Potential Viral Vaccine Needs

In this brief examination of vaccine candidates, I have obviously been selective, for many other vaccines are in the works. Furthermore, I have completely ignored the group of pathogens where cell substrates are the most markedly rate-limiting: hepatitis A, non-A-non-B hepatitis, JC papovavirus, human rotaviruses, and a host of others are still at "square one" because of a lack of any suitable cell substrate; and success at their cultivation might well yield pathogenetic data which would make vaccine development desirable and feasible. My initial survey of basic science advances pertains doubly to these important areas of future virologic advance but can only receive passing notice in this presentation.

SUMMARY REMARKS

In summary then, my purpose has been two-fold: on the one hand, I have tried to highlight the kinds of basic science advances in both cellular and virologic research that can (and should) be focussed both on vaccines under development and, retrospectively, on those whose origins were strictly empiric. On the other hand, I have attempted a partial survey of some of the prominent members of a potential new generation of vaccines to point out areas where these advances can and should contribute either to progress or to a sense of caution about the further reliance on pure empiricism.

It is clear that we are not finished with new viral vaccines. It is equally clear that narrowing the persistent gap between basic science and its application to public health needs will require much energy and attention as vaccine development progresses.

BIBLIOGRAPHY

1. Meyer, H.M. Jr., H.E. Hopps, P.D. Parkman, et al. Control of measles and rubella through use of attenuated vaccines. Amer. J. Clin. Path. 70:128. 1978.
2. Madoff, M.A., Gleckman, R. A. Immunizations: where the money should be. J. Inf. Dis. 133:230. 1976.
3. Sabin, A.B. Overview and horizons in prevention of some human infectious diseases by vaccination. Amer. J. Clin. Path. 70: 114. 1978.
4. Melnick, J.L. Viral vaccines: new problems and prospects. Hospital Practice, p. 104, July 1978.
5. Protection of human subjects: research involving children. Federal Register 43:31786, 1978.
6. Hofmann, F. The growing federal presence in human investigative studies. Fed. Proc. 36:133. 1977.
7. Robbins, F.C. Demand for human trials in biological research. Bull. W.H.O. 55:73. 1977.
8. Jonsen, A. Research involving children: recommendations of the National Commission for the Protection of Human Subjects of Biomedical and Behavioral Research. Pediatrics. 62:131. 1978.
9. Holden, C. Animal rights advocate urges new deal. Science. 201:35. 1978.
10. Wade, N. New vaccine may bring man and chimpanzee into tragic conflict. Science. 200:1027. 1978.
11. Slate, D.L., Shulman, L., Lawrence, J.B. et al. Presence of human chromosome 21 alone is sufficient for hybrid cell sensitivity to human interferon. J. Virol. 25:319. 1978.
12. Crowell, R.L. Specific viral interference in HeLa cell cultures chronically infected with Coxsackie B5 virus. J. Bact. 86: 517. 1963.
13. Fareed, G., Davoli, D. Molecular biology of papovaviruses. Ann. Rev. Biochem. 46:471 1977.
14. St. Jeor, S.C., Hutt, R. Cell DNA replication as a function in the synthesis of human cytomegalovirus. J. Gen. Virol. 37: 65. 1977.
15. DeMarchi, J.N., Kaplan, A.S. Replication of human cytomegalovirus DNA: lack of dependence on cell DNA synthesis. J. Virol. 18:1063. 1976.
16. DeMarchi, J.N., Kaplan, A.S. Physiologic state of human embryonic lung cell affects their response to human cytomegalovirus. J. Virol. 23:126. 1977.
17. Temin, H.M., Baltimore, D. RNA-directed DNA synthesis and RNA tumor viruses. Adv. Virus Res. 17:129. 1972.
18. St. Jeor, S.C., Rapp, F. Cytomegalovirus: conversion of nonpermissive cells to a permissive state for virus replication. Science 181:1060. 1973.
19. Arber, W., Linn, S. DNA modification and restriction. Ann. Rev. Biochem. 36:467. 1969.

20. Meselson, M., Yuan, R., Heywood, J. Restriction and modification of DNA. Ann. Rev. Biochem. 41:447. 1972.
21. Ishida, N., Homma, M. A variant Sendai virus, infectious to egg embryos but not to L cells. Tohoku J. Exp. Med. 73: 56. 1960
22. Scheid, A., Choppin, P.W. Identification of biological activities of paramyxovirus glycoproteins. Activation of cell fusion, hemolysis and infectivity by proteolytic cleavage of an inactive precursor protein of Sendai virus. Virology. 57:475. 1974.
23. Scheid, A., Choppin, P.W. Two disulfide-linked polypeptide chains constitute the active F protein of paramyxoviruses. Virology. 80:42. 1977.
24. Osborn, J.E., Walker, D.L. Virulence and attenuation of murine cytomegalovirus. Infect. Immun. 3:228. 1971.
25. Mannini, A., Medearis, D.N. Jr. Mouse salivary gland virus infections. Amer. J. Hyg. 73:320. 1961.
26. Osborn, J.E., Medearis, D.N. Jr. Studies of the relationship between mouse cytomegalovirus and interferon. Proc. Soc. Exp. Biol. Med. 121:819. 1966.
27. Osborn, J.E., Blazkovec, A.A., Walker, D.L. Immunosuppression during acute murine cytomegalovirus infection. J. Immunol. 100: 835. 1968.
28. Howard, R.J., Balfour, H.H. Jr., Seidel, M.V. et al. Effect of murine cytomegalovirus on cell-mediated immunity. Transplant. Proc. 9:355. 1977.
29. Reddy, V.B., Thimmappaya, B., Dhar, R. et al. The genome of simian virus 40. Science. 200:494. 1978.
30. Schaffer, P.A. Temperature-sensitive mutants of herpesvirus. Current Topics in Microbiol. and Immunol. 70:51. 1975.
31a. Pereira, L., Cassai, E., Honess, R.W. et al. Variability in the structural polypeptides of herpes simplex virus 1 strains: potential application in molecular epidemiology. Infect. Immun. 13:211. 1976.
31b. Buchman, T.G., Roizman, B., Adams, G. et al. Restriction endonuclease fingerprinting of herpes simplex virus DNA: a novel epidemiological tool applied to a nosocomial outbreak.
31c. Milstein, J.B., Walker, J.R., Eron, L.J. Correlation of virus polypeptide structure with attenuation of poliovirus type 1. J. Virol. 23:811. 1977.
32. Ashe, W.K., Notkins, A.L. Neutralization of an infectious herpes simplex virus-antibody complex by anti-gamma globulin. Proc. Nat. Acad. Sci. U.S.A. 56:447. 1966.
33. Graham, F.L., Veldhuisen, G., Wilkie, N.M. Infectious herpesvirus DNA. Nature New Biol. 245:265. 1973.
34. Geelan, J.L.M.C., Walig, C., Werthein, P. et al. Human cytomegalovirus DNA. I. Molecular weight and infectivity. J. Virol. 26:813. 1978.

35. Schmidt, N.J., Dennis, J., Lennette, E.H. Complement-fixing reactivity of varicella-zoster virus subunit antigens with sera from homotypic infections and heterotypic herpes simplex virus infections. Infect. Immun. 15:850, 1977.
36. Palese, P., Ritchey, M.B. Live attenuated influenza virus vaccine strains with temperature-sensitive defects in P3 protein and nucleoprotein. Virology. 78:183. 1977.
37. Richman, D.D., Murphy, B.R., Chanock, R.M. Demonstration of a non-temperature-sensitive growth-restriction mutation in a ts mutant of influenza A virus: implications for live virus vaccine development. Virology. 83:356. 1977.
38. Douglas, R.G., Betts, R. Studies with live attenuated influenza vaccines. Amer. J. Clin. Path. 70:153. 1978.
39. Moss, B., Keith, J.M., Gerhoswitz, A., Ritchey, M.B., Palese, P. Common sequence at the 5' ends of the segmented RNA genomes of influenza A and B viruses. J. Virol. 25:312. 1978.
40. Meyer, H.M. Jr., Hopps, H.E., Parkman, P.D. et al. Review of existing vaccines for influenza. Amer. J. Clin. Path. 70:146. 1978.
41. Tyeryar, F.J., Richardson, L., Belshe, R. Report of a workshop on respiratory syncytial virus and parainfluenza viruses. J. Inf. Dis. 137:835. 1978.
42. Kim, H.W., Leikin, S.L., Arrobio, J., et al. Cell-mediated immunity to respiratory syncytial virus induced by inactivated vaccine or by infection. Pediatric Res. 10:74. 1976.
43. Craighead, J.E. Report of a workshop: disease accentuation after immunization with inactivated microbial vaccines. J. Inf. Dis. 131:749. 1975.
44. Chanock, R.M., Richardson, L.S., Belshe, R.B. et al. Prospects for prevention of bronchiolitis caused by respiratory syncytial virus. Pediatric Res. 11:264. 1977.
45. Chanock, R.M., Kim, H.W., Brandt, C. et al. Respiratory syncytial virus, in Viral Infections of Humans: Epidemiology and Control. (A.S. Evans, ed.) Plenum Medical Book Co., New York. 1976.
46. Bruhn, F.W., Yeager. A.S. Respiratory syncytial virus in early infancy: circulating antibody and the severity of infection. Amer. J. Dis. Child. 131:145. 1977.
47. McIntosh, K., Master, H.B., Orr, I. et al. The immunologic response to infection with respiratory syncytial virus in infants. J. Inf. Dis. 138:24. 1978.
48. Buynak, E.B., Weibel, R.E., McClean, A.A. et al. Live respiratory syncytial virus vaccine administered parenterally. Proc. Soc. Exp. Biol. Med. 157:636. 1978.
49. Maupas, P., Goudeau, A., Coursaget, P. et al. Immunisation against hepatitis B in man. Lancet i:1367. 1976.
50. Robinson, W.S., Lutwick, L. The virus of hepatitis type B. New Eng. J. Med. 295:1168. 1976.
51. Robinson, W.S. The genome of hepatitis B virus. Ann. Rev. Micro. 31:357. 1976.

52. Buynak, E.B., Roehm, R.R., Tytell, A.A. et al. Vaccine against human hepatitis B. J.A.M.A. 235:2832. 1976.
53. Purcell, R.H., Gerin, J.L. Hepatitis B vaccines: on the threshold. Amer. J. Clin. Path. 70:159. 1978.
54. Zuckerman, A.J. Oncogenic potential of new viral vaccines. Nature 272:579. 1978.
55. Maynard, J.E., Berquist, K.R., Krushak, D.H. Experimental infection of chimpanzees with the virus of hepatitis B. Nature. 237: 514. 1972.
56. Barker, L.F., Chisari, F.V., McGrath, P.P. Transmission of type B viral hepatitis to chimpanzees. J. Inf. Dis. 127: 648. 1973.
57. Gershon, A.A. Varicella-zoster virus: prospects for active immunization. Amer. J. Clin. Path. 70:170. 1978.
58. Brunell, P.A., Wolman, S.R., Steinberg, S. Propagation of varicella-zoster virus in a diploid strain of embryonic thyroid cells from the rhesus monkey. J. Inf. Dis. 125:545. 1972.
59. Caunt, A.E. Growth of varicella-zoster in human thyroid tissue cultures. Lancet ii:982. 1963.
60. Brunell, P.A. Separation of infectious varicella-zoster virus from human embryonic lung fibroblasts. Virology 31:732. 1967.
61. Rifkind, D. The activation of varicella-zoster virus infections by immunosuppressive therapy. J. Lab. Clin. Med. 68: 463. 1966.
62. Sabin, A.B. Varicella-zoster vaccine. J.A.M.A. 238:1731. 1977.
63. Shibuta, H., Ishikawa, T., Hondo, R. et al. Varicella virus isolation from spinal ganglion. Arch. f.d. Ges. Virusforsch. 45:382. 1974.
64. Kempe, C.H., Gershon, A.A. Varicella vaccine at the crossroads. Pediatrics 60:930. 1977.
65. Asano, Y., Nakayama, H., Yazaki, T. et al. Protection against varicella in family contacts by immediate inoculation with live varicella vaccine. Pediatrics. 59:3. 1977.
66. Asano, Y., Takahashi, M. Clinical and serologic testing of a live varicella vaccine and two year follow-up for immunity of the vaccinated children. Pediatrics 60:810. 1977.
67. Asano, Y., Nakayama, H., Yazaki, T. et al. Protective efficacy of vaccination in children in four episodes of natural varicella and zoster in the ward. Pediatrics 59:8. 1977.
68. Kamiya, H., Ihara, T., Hattori, A. et al. Diagnostic skin test reactions with varicella virus antigen and clinical application of the test. J. Inf. Dis. 136:784. 1977.
69. Baba, K., Yabuuchi, H., Okuni, H. et al. Studies with live varicella vaccine and inactivated skin test antigen: protective effect of the vaccine and clinical application of the skin test. Pediatrics 61:550. 1978.
70. Izawa, T., Ihara, T., Hattori, A. et al. Application of a live varicella vaccine in children with acute leukemia or other malignant diseases. Pediatrics 60:805. 1977.

71. Stevens, D.A., Merigan, T.C. Interferon, antibody and other host factors in herpes zoster. J. Clin. Invest. 51:1170. 1972.
72. Elek, S.D., Stern, H. Development of a vaccine against mental retardation caused by cytomegalovirus infection in utero. Lancet 1:1. 1974.
73. Just, M., Buergin-Wolff, A., Emoedi, G., et al. Immunisation trials with live, attenuated cytomegalovirus (Towne 125). Infection 3: 111. 1975.
74. Plotkin, S.A., Farguhar, J., Hornberger, E. Clinical trials of immunization with the Towne 125 strain of human cytomegalovirus. J. Inf. Dis. 134:470. 1976.
75. Weller, T.H. The cytomegaloviruses: ubiquitous agents with protean clinical manifestations. New Eng. J. Med. 285:203, 267. 1971.
76. Osborn, J.E. Diseases caused by human cytomegaloviruses. Chapter 37C in Brennemann-Kelley's Textbook of Pediatrics, Hoeber Medical Division, Harper and Rowe, New York. 1975.
77. Huang, E.S., Kilpatrick, B.A., Huang, T. et al. Detection of human cytomegalovirus and analysis of strain variation. Yale J. Biol. Med. 49:29. 1976.
78. Zablotney, S.L., Wentworth, B.B., Alexander, E.R. Antigenic relatedness of 17 strains of human cytomegalovirus. Amer. J. Epidemiol. 107:336. 1978.
79. Beutner, K.R., Morag, A., Deibel, R. et al. Strain-specific local and systemic cell-mediated immune responses to cytomegalovirus in humans. Infect. Immun. 20:82. 1978.
80. Waner, J., Weller, T.H. Analysis of antigenic diversity among human cytomegaloviruses by kinetic neutralization tests with high-titered rabbit antiser. Infect. Immun. 21:151. 1978.
81. Waner, J., Weller, T.H., Kevy, S. Patterns of cytomegaloviral complement-fixing antibody activity: a longitudinal study of blood donors. J. Inf. Dis. 127:538. 1973.
82. Gold, E., Nankervis, G.A. Cytomegalovirus. Chapter 7 in Viral Infections on Humans: Epidemiology and Control. (A.S. Evans, Ed.) Plenum Medical Book Co., New York. 1976.
83. Rola-Pleszczynski, M., Frenkel, L.D., Fuccillo, D.A. et al. Specific impairment of cell-mediated immunity in mothers of infants with congenital infection due to cytomegalovirus. J. Inf. Dis. 135:386. 1977.
84. Gehrz, R.C., Marker, S.C., Knoff, S.O. et al. Specific cell-mediated immune defect in active cytomegalovirus infection of young children and their mothers. Lancet ii:844. 1977.
85. Yaeger, A.S., Martin, H.P., Stewart, J.A. Congenital cytomegalovirus infection: outcome for the subsequent sibling. Clin. Pediatrics. 16: 455. 1977.
86. Stagno, S., Reymolds, D.W., Huang, E.-S., et al. Congenital cytomegalovirus infection: occurrance in an immune population. New Eng. J. Med. 296:254. 1977.

87. Reynolds, D.W., Stagno, S., Hosty, T.S. et al. Maternal cytomegalovirus excretion and perinatal infection. New Eng. J. Med. 289:1. 1973.
88. Reynolds, D.W., Stagno, S., Reynolds, R. et al. Perinatal cytomegalovirus infection: influence of placentally transferred maternal antibody. J. Inf. Dis. 137:564. 1978.
89. Nankervis, G.A., Kumar, M.L., Gold, E. Primary infection with cytomegalovirus during pregnancy. Ped. Res. 8:427, 1974.
90. Howard, R.J., Balfour, H.H. Jr. Prevention of morbidity and mortality of wild murine cytomegalovirus by vaccination with attenuated cytomegalovirus. Proc. Soc. Exp. Biol. Med. 156: 365. 1977.
91. Medearis, D.N. Jr., Prokay, S.L. Effect of immunization of mothers on cytomegalovirus infection in suckling mice. Proc. Soc. Exp. Biol. Med. 157:523. 1978.
92. Osborn, J.E., Walker, D.L. Virulence and attenuation of murine cytomegalovirus. Infect. Immun. 3:228. 1970.
93. Hsiung, G.D., Choi, Y., Bia, F. Cytomegalovirus infection in quignea pigs. I. Viremia during acute primary and chronic persistent infection. J. Inf. Dis. 138:191. 1978.
94. Choi, Y., Hsiung, G.D. Cytomegalovirus infection in guinea pigs. II. Transplacental and horizontal transmission. J. Inf. Dis. 138:197. 1978.
95. Hanshaw, J.B. A cytomegalovirus vaccine? Amer. J. Dis. Child. 128:141. 1974.
96. Medearis, D.N. Jr. Human cytomegalovirus immunization prospects. New Eng. J. Med. 296:1289. 1977.
97. Plotkin, S.A., Farquhar, J., Hornberger, E. Clinical trials of immunization with the Towne 125 strain of human cytomegalovirus. J. Inf. Dis. 134:470. 1976.
98. Stern, H. Cytomegalovirus infection in the neonate and its prevention. Postgrad. Med. J. 53:588. 1977.
99. Glazer, J.P., Friedman, H.M., Grossman, R.A. et al. Cytomegalovirus vaccination and renal transplantation. Lancet i: 90. 1978.
100. Babiuk, L.A., Rouse, B.T. Immune interferon production by lymphoid cells: role in the inhibition of herpesviruses. Infect. Immun. 13:1567. 1976.
101. Kitces, E.N., Morahan, P.S., Tew, J.G. et al. Protection from oral herpes simplex virus infection by a nucleic acid-free virus vaccine. Infect. Immun. 16:955. 1977.
102. Wise, T.G., Pavan, P.R., Ennis, F.A. Herpes simplex virus vaccines. J. Inf. Dis. 136:706. 1977.
103. Parks, W.P., Rapp, F. Prospects for herpesvirus vaccination - safety and efficacy considerations. Progr. Med. Virol. 21: 188. 1975.

CHOICE OF CELL SUBSTRATE FOR BIOLOGICAL PRODUCTS

A. J. Beale

The Wellcome Research Laboratories

Beckenham, Kent, England

The methods at present adopted for the control of the substrate for production of modified live virus vaccines are well established. They involve maintaining colonies of chickens, ducks or rabbits which are the source of primary cells for making such vaccines. These colonies are regularly monitored to exclude infections. The use of human diploid cell lines like WI38 and MRC5 are now internationally accepted as satisfactory substrates for these vaccines and bring the benefits of a defined seed system to the cell substrates as well as to the virus seed. These benefits are consistency and stability; and the ability to use a pretested seed.

There is a problem for some other important biological products derived from animals or cells in culture, for example, hormones and killed virus vaccines. It is often difficult to obtain products like insulin, or interferon, or killed rabies vaccine in sufficient quantities at a reasonable cost with existing techniques, and so alternative methods have to be considered. At present synthesis is not possible, and the production of these substances in bacteria by the techniques of genetic manipulation is not yet a practical reality, although for insulin and perhaps for interferon they may not be too far distant.

Consideration will therefore be given to the possible use for these purposes of cells that can be produced in mass culture. Thus, for example, baby hamster kidney cells (BHK 21 clone 13) have been used extensively for the manufacture of killed foot and mouth disease vaccine (1) and Newcastle Disease Vaccine (2). These and other similar cells are "transformed" and some are tumorigenic and suspected of oncogenic potential. Is it possible to use such cells for production of biological products destined for man? They have in the past been suggested for example, by an influential N.I.H. Committee in 1967(3), but have not so far achieved general acceptance by the scientific community.

When faced with this sort of question guidance should be taken from precedent and from the availability of alternatives. Also, there are steps that can be taken to identify and reduce the potential hazards, and these should be taken into account. Finally, the method of use in medicine is crucial in determining the risks and benefits.

In the search for precedent, one can first consider some sources currently used to produce biological products. Potentially or actually contaminated materials have been and are being used as the source of products for human medicine. These are illustrated by the following examples:

i) Antibodies are obtained from the blood of animals or man. Clearly man is a better source because of the serum sickness associated with the injection of foreign serum (4). Also, human blood has proved a satisfactory source of a number of other products, in addition to specific gamma globulins. Nevertheless, blood and serum are potentially contaminated with a number of agents, the most serious being hepatitis viruses. Despite this, products have been successfully prepared from this source because the contaminating agents are destroyed by the purification and inactivation procedures used for most of the products. It should be noted that destruction of hepatitis B viruses has been achieved empirically. Before this was certain, risks were taken with such products, because the benefits which resulted from their use were considered to outweigh the risks. Looking ahead it should be possible to produce essentially pure monoclonal antibodies of a desired specificity by hybridising antibody-forming cells with a cell capable of growing in culture, for example, a lymphoblastoid cell induced to proliferate continuously by EB virus. Will these be acceptable for medical purposes?

ii) Insulin is prepared from bovine and porcine pancreas, collected on a world wide scale from abattoirs under conditions where indigenous and exogenous contamination with microbial agents must be rife. The extraction and purification procedures, however, yield very highly purified and sterilized hormone.

iii) Inactivated influenza vaccines are made from commercial fertile hens eggs, which are known to be contaminated by a variety of avian viruses, including leukosis viruses. The viruses are destroyed by the formalin or beta propiolactone used during vaccine manufacture. Moreover, influenza virus vaccines are nowadays purified so that they are essentially biochemically homogenous virus preparations; or the isolated haemagglutinin and neuraminadase viral surface antigens may be separated from whole virus and used as vaccine.

iv) Killed poliovaccine is still mainly made from primary cultures of monkey kidney cells, and these are notoriously contaminated with extraneous agents (5), particularly Simian viruses. They have also been shown to contain reverse transcriptase for oncornaviruses. One large manufacturer of living poliovaccine has reported that about 50% of primary monkey kidney cultures had to be rejected because of viral contamination (6). With such a rate of contamination it seems certain that some contaminated culture will in fact have been used to make vaccine batches. Indeed batches were found to contain SV40 including a highly purified vaccine. Monkey cytomegalovirus and foamy viruses are frequent contaminants. Nevertheless, as the result of inactivation and a relatively small amount of purification, a product has been obtained which has been shown to be effective and safe. Moreover, improved killed poliovaccine has been very successful in the control of poliomyelitis in Sweden, Holland, Denmark and some provinces of Canada. It has also been extensively used in France.

V) Most of the human interferon that has been used so far in patients has been made from buffy coats obtained from blood donations. These donations are from donors screened especially in regard to hepatitis B. The buffy coat cells will in many instances contain Epstein Barr virus (EB) or cytomegalovirus (CMV) and no doubt other agents from time to time. The interferon is however processed to yield a sterile product.

These examples serve to show that there is ample precedent for making biological products and vaccines from a source which is either actually constaminated or potentially contaminated with microbial agents, including some which may be oncogenic. After purification and with chemical treatment, safe and effective products are obtained.

Illustrative examples of the problems of the choice of cell substrate to be considered here are the production of interferon and of killed poliovaccine. In order to make large amounts of human interferon, there are various possible cell sources as shown in Table 1. Buffy coat cells have so far been used to produce nearly all the interferon used in man. However, it will be extremely difficult to produce on a really large scale a consistent commercial product from such a source. One blood donation provides about 1 mega unit of interferon, and in Finland, some 1.8×10^5 mega units per annum have been produced (Cantell personal communication) an amount enough to treat some hundreds of cases of hepatitis or osteogenic sarcoma. Moreover, human buffy coat cells are a primary cell substrate which can never be fully controlled. The cells may be contaminated with agents such as Epstein Barr virus and cytomegalovirus,

Table 1

Available Cells for Production of Human Interferon
Buffy Coat Cells
Fibroblast Cell Strains
Lymphoblastoid Cells

which will vary with the batch, and the risks are likely to increase as the number of donors and the scale of the operation increases.

Human fibroblast diploid cell lines have also been used as a source of interferon, but the molecule is different in its chemical and biological properties from human leukocyte interferon, and should be considered as an alternative rather than a substitute for the latter. In addition, there are problems of stability *in vitro* and *in vivo* with fibroblast interferon. Fibroblast cells are anchorage dependent and the scaling up of production of such cells is technologically difficult. At the Wellcome Research Laboratories, for these sorts of reasons, we have decided to make interferon from human lymphoblastoid cells. Lymphoblastoid interferon seems closely similar to, if not identical with, human leukocyte interferon, the main difference being a higher content of fibroblast interferon. Among lymphoblastoid cells we have chosen the Namalwa line for a number of reasons. These cells are among the highest yielders of interferon out of many cell lines which we have tested (7), now more than 170 in number, and have good growth properties. Although they are derived from a Burkitt tumour patient, and they contain part of the EB virus genome, no infectious virus is formed either spontaneously, or in response to treatment with BUdr. Indeed the cells contain only some of 50% of the EB virus genome, with two such genome copies per cell (8). Of the DNA sequences present, only three are translated into polyadenylated RNA that becomes associated with polysomes (Kieff Personal Communication). One of these genes must code for the EB nuclear antigen (EBNA) found in this, as in all other human lymphoblastoid cell lines. The EB Early Antigen and Virus Capsid Antigen are not formed, confirming absence of a productive infection.

When used in production, the Namalwa cells can be handled according to the accepted principles of a seed management system for vaccine production. Thus a Master Cell Seed bank has been established and has been tested for microbial contaminants. They have also been tested for tumorigenicity in athymic mice according to established procedures and found to be negative. It must be recorded however that some tumorogenicity has been reported in a more prolonged test by others (9). The cells are heteroploid, but they can be characterized karylogically, as a guide to contamination or to a major spontaneous mutation: over a two year period there has been no significant change in karyotype. The cells are used in production during a three month period, but time is also required to build up the necessary number of cells from a frozen ampoule of Master Cell Seed. Thus a total of some 75 subcultures are involved.

The cells are grown in suspension culture in large vessels, at present up to 1000 litres, but based on experience with foot and mouth disease vaccine in BHK cells, there seems no reason why larger vessels could not be used. The interferon is induced using Sendai virus and then purified. At the Wellcome Research Laboratories methods involving antibody affinity chromatography and trichloro-acetic acid precipitation and ethanol extraction for purification and inactivation steps have been developed (10). We have pioneered the use of antibody affinity columns for the purification of biological products, namely tetanus (11) and cholera toxoid and they yield highly purified products, uncontaminated with significant amounts antibody which might leak from the columns judged by antibody production in man.

Interferon produced in this way routinely has a purity of more than 10^7 units per milligram of protein. The materials that might contaminate the interferon are calf serum used for growth of cells, Namalwa cell proteins and DNA including EB viral DNA and Sendai virus and allantoic fluid.

The results for calf serum which is present in large quantities are shown in Table 2. It is substantially removed by the processing. Three methods have been used for estimation of residual bovine serum content: (i) immunoelectrophoresis, (ii) challenging sensitised guinea pigs, and (iii) adding ^{125}Iodine labelled serum to the interferon being processed (Table 3). More recent experience with a better production scale column suggests that even less 'calf serum' is detectable in the final product.

The effect of the purification and inactivation steps have been assessed against a number of viruses and nucleic acids. The interferon antibody affinity column which is used is highly specific, and antibodies against contaminants in the antigen preparation used to raise the serum are removed by adsorption, for example, with Sendai virus and egg protein in order to purify the specific globulin before they are coupled to form a column. In order to avoid problems with antibodies to calf serum the antisera were raised by immunizing calves. Most viruses that have been studied are not retained in the "scrubbed" antibody affinity column and in most instances, any bound virus is destroyed by the low pH of the eluting fluid. Thus, for example, no Sendai virus is found in the column eluates. Similarly, when radio-labelled EBV DNA or whole Raja cell DNA was taken through the column (these were kindly supplied by Dr. J. Pagano), no detecta-

Table 2

Purification of Interferon based on total and ^{125}I calf serum protein

Preparation	Total protein units/mg	Total protein purification factor	^{125}I calf serum protein IF/cpm	^{125}I calf serum protein purification factor
Initial	1.4×10^4	1	0.69	1
Purified	6.6×10^7	4500	35,000	51,000

Table 3

Calf Serum Content of Purified Interferon

Method	per mega unit
Radioactivity	1.25 micrograms*
Guinea pig sensitivity	7-70 ng
Counter current immunoelectrophoresis	27 ng**

*An overestimate because radioactivity not significantly retained by:

(a) Cibacron blue or
(b) on an anticalf serum column

**Results from A. van Wezel

ble counts and 0.0068% of the original counts were detected in the column eluate respectively (Table 4). Even these small amounts were lost on further purification with TCA ethanol. We have also checked the effectiveness of the TCA/ethanol step at inactivating viruses. In early studies, animal viruses were added to crude interferon which had not been through an antibody column and processed with TCA/ethanol. However, the very high concentration of human interferon often caused some interference with the growth of these viruses even in nonhuman cells, and only with avian viruses, could satisfactory data be obtained. Therefore, in later studies, interferon was added to a maintenance medium with serum added to match the protein content normally obtained from the column eluate (Table 5). In such tests, it has been shown that feline rhinotracheitis virus and poliovirus Type 1 are completely inactivated. In a larger series of experiments designed to overcome these technical problems a range of bacteriophages differing in size and shape, and in the nature of their nucleic acid, have been processed, and all were successfully destroyed (Table 6).

In the experiment with radio-labelled Raji cellular DNA, only 6.7×10^{-7}% of the initial radioactivity was found in the final fraction, corresponding to purified interferon. Presumably the figure would have been similar for Namalwa cell DNA. It can be calculated that only 0.25×10^{-4} of the total DNA obtained from Namalwa cells is EBV DNA (Kieff Personal Communication). On this basis, the residual EBV DNA in purified interferon will be of the order of 1.7×10^{-11} of the original very low level in the crude interferon.

In summary therefore we have used lymphoblastoid cells because of all the possibilities they seem most likely to yield the best interferon free from oncogenic potential. Moreover it can be produced efficiently and economically in sufficient quantity. Selected cells are used and a semi-automatic enclosed production system employed, followed by a series of purification steps - immunopurification, TCA/ethanol and filtration each of which reduces the level of possible contaminating materials or infectious agents, and which taken together give a high probability of yielding a safe product for use in medicine. The nature of this use is perhaps also pertinent. In the first instance interferon will be used in trials involving patients with malignant disease or severe virus infections or both. As safety and effectiveness is established, the materials from this source can then be used with increasing confidence for other medical indications.

A second example of the problems of choice is the substrate for killed poliovirus production. It is clear that killed polio vaccine can be very effective in developed countries, like Sweden,

Table 4

Removal of EB virus DNA during purification of Interferon

Sample	Description	Effective vol (ml)	cpm/ml (corrected)	Total cpm (corrected)	%cpm
1.	Starting material	5,000	77.1	385,000	100
2.	Column filtrate	5,000	80.5	402,500	105
3.	Saline wash	500	7.2	6,100	1.6
4.	Citrate eluate	350	76.1	2,300	3.2
5.	Final ("purified interferon")	5	-1.3	6.5	-0.002

Table 5

Destruction of representative virus by TCA/ethanol

Agent	Diluted in	Infectivity initial (log/ml)	Final dilution tested	Maximum residual infectivity (%)
1. Interferon (possibly) active in test system				
Canine hepatitis	IF	5.7	10^{0} :none found	2×10^{-6}
Mouse CMV	IF	5.2	10^{-1} :none found	6×10^{-5}
Polyoma	IF	6.2	10^{-1} :none found	1×10^{-5}
Polyoma DNA	IF	2.7	10^{-1} :none found	2×10^{-2}
2. Interferon inactive in test system				
NDV	IF	7.55	10^{0} :none	none
		7.3	10^{0} :none	none
Sendai	IF	6.15	10^{0} :none	none
3. Virus diluted in serum-containing medium				
Polio Type 1		6.4	10^{0} :none found	4×10^{-7}
Cat herpes		3.8	10^{-1}:none found	1×10^{-4}
Sendai		8.0	10^{0} :none found	

Virus added to 1000 ml interferon or diluent, and concentrated 100 times.

IF = Interferon

Table 6

Destruction of the infectivity of bacteriophages representative of different classes by TCA/Ethanol

Bacteriophage	NA type	Before treatment	After treatment
T3	ds-DNA	1.7×10^{9}	none detected
W31	ds-DNA	2.0×10^{8}	none detected
BF23	ds-DNA	1.2×10^{11}	none detected
ϕX174	ss-DNA	1.4×10^{11}	none detected
1-fil	ss-DNA	7.5×10^{5}	none detected
MS2	ss-RNA	1.3×10^{10}	none detected

*20-fold concentration by volume

ds = double stranded

ss = single stranded

Finland, Holland and Canada. It could also solve the problem of poor results with OPV vaccination in the tropics. Using the D antigen system for potency testing described in 1961 (12), it has been found that a single dose of killed poliovaccine with a D antigen content of Type 1 80 units, Type 2 8 units and Type 3 64 units produces an antibody response in virtually all subjects in the tropics (13).

The problem is, can killed poliovaccine be prepared on an adequate scale and economically to meet the requirements of the developed and developing world. The available alternative cell substrates are monkey kidney cells; human diploid cell lines, or a transformed cell line capable of growing in suspension. If the yield of virus is considered, it is found that a transfomed cell line is much m('e efficient at production in that it yields more D antigen per millilitre (Table 7). Clearly a systematic search for the optimal virus and cell type is required. Some of the aspects of choice are shown in Table 8.

There is room for some argument about the relative merits of surface adherent and suspension culture, but the lessons of antibiotic production will make the industrial technologist opt for suspension culture as the most promising for scale up and technological improvement. MRC5 cells or other human diploid cell lines seem an inefficient system, yielding low titres and making concentration and purification difficult and costly. The consequences of this use of inefficient technology is eloquently revealed by the shortage and high price of killed rabies vaccine made in those cells. Moreover, the reasons for using human diploid cells rather than the more efficient primary dog kidney (14) or BHK cells are far from compelling.

Monkey kidney cells also seem a poor option on first consideration. There is an every more threatening shortage of wild caught animals, and breeding animals for production as opposed to testing seems a formidable undertaking. Moreover, the frequent contamination of these cells is an added deterrent which does not make them a favored substrate. There are some positive aspects. Poliovaccines, both living and killed have been used extensively, successfully and safely, when prepared on this substrate, so that its safety is abundantly proved. van Wezel (14) has developed an improved technique of production of killed viral vaccines using monkey kidney cells attached to Sephadex beads that greatly facilitates the process from a standpoint of industrial technology. He estimates 150 monkeys are required to make Dutch formula vaccine, i.e. Type 1: 40 D units; Type 2: 4 D units; Type 3: 8 units. Moreover, "second generation" cells which were accepted by the British Control Authorities in the sixties would enable more vaccine to be made from the cells. Indeed, as few as 25 monkeys may be needed for a million doses of vaccine making the production of vaccine from home bred

Table 7

Yield of poliovirus in D units/ml from various cell substrates

	Cell lines (Hela)	Primary Monkey kidney	Diploid cell lines (MRC 5)
Poliovirus			
Type 1	350	80	10
Type 2	200	25	10
Type 3	200	35	50

Table 8

Aspects of choice of cell substrate

	Cell lines	Monkey kidney	Diploid cell lines
Availability	Unlimited	Shortage of monkeys Breeding possible	Limited only by finite life
Yield	High	High	Low
Culture Technology	Suspension culture scale up easy	Microcarrier scale up possible	Anchorage dependent microcarrier ±
Potential Contamination	Transformed	Many contaminants Bred animals cleaner	Clean

animals feasible: moreover they contain fewer viruses particularly less foamy virus contamination (14). Production of vaccine may be possible for some countries but it is open to doubt whether the needs of the world for poliovaccine can be met in this way.

As referred to earlier, a continuous cell line has been used for foot and mouth disease vaccine production, and the living cells themselves have not been found to produce tumors when deliberately injected into cattle, pigs, dogs, cats and poultry. Moreover, no product from these cells has been shown to be oncogenic in animals: thus even minimal treatment is sufficient to destroy their tumourgenic activity assessed in the hamster cheek pouch (2), and several such steps are employed, e.g. viral infection, filtration, formalin treatment, viral purification; since each singly is adequate the cumulative safety factor is very great.

Hundreds of millions of doses of vaccines made from BHK 21 cells have now been given to animals. Although most of these animals have been slaughtered for food purposes, some groups of cattle and poultry have been kept for their full life span, and no ill effects attributable to the vaccine have been observed.

The United States Department of Agriculture seems to have encouraged the use of cell lines rather than primary cells for veterinary virus vaccine production, provided the cells do not produce tumors in the target species for the vaccine, because they provide tighter control and allow for a seed test system.

To revert to poliovaccine, the important steps in its preparation are the purification, filtration and inactivation. These are of course similar for all methods of producing the raw virus fluid, but the higher the initial titre the more efficient such steps are, and the more economically they can be achieved.

van Wezel (14) has described a method for purifying and concentrating poliovirus which involves first concentration by hollow fibre ultrafiltration with a cut off of 100,000 mol. wt. It is subsequently purified by gel filtration through a Sepharose 6B column and finally passed through a DEAE Sephadex column.

It will cause no surprise that at Wellcome we have used an antibody affinity column in addition to DEAE Sephadex columns. The antibody affinity column provides an additional type of purification procedure which is always advantageous for ensuring the removal of contaminants. Highly specific antisera are available for poliovirus and, for example, will not absorb closely related viruses like that of Swine Vesicular Disease or ^{35}S methionine labelled cellular material.

In summary it is now possible technologically to produce essentially pure poliovirus, for example, giving very clean polyacrilamide gel preparations, on a large scale. This is probably the most efficient way, and may be the only practical way for meeting the needs of the world for killed poliovaccine supply if that vaccine becomes the method of choice for the control of poliomyelitis. Such a vaccine would still contain viral nucleic acid, although treated with formalin which is clearly an adequate treatment for virus produced on monkey kidney cells. It is possible to dream of the eradication of poliomyelitis and this dream might become a reality if the best technology of production were used. Theoretically a vaccine containing only the nucleic acid free immunogen would be preferred.

Many years ago Pereira (15, 16) produced cystalline adenoviruses type 5 hexons and fibres, and these are ideal vaccines enabling the antigen content to be controlled by weight (17).

Clearly, the best means for producing them would be in a transformed line also. A vaccine for rabies containing only subunits could also have attractions but so far at least such preparations have a lesser immunogenicity.

REFERENCES

1. Capstick P.B., Garland A.J., Chapman W.G., et al. Production of foot and mouth disease virus antigen from BHK 21 clone 13 cells grown and infected in Deep Suspension Cultures. Nature 205:1135-1136, 1965.
2. Keeble S.A. and Heymann C.S.. Effect of β-Propiolactone and Thiomersal on Growth of B.H.K. cells in the Hamster Cheek Pouch. Nature 208:1125-1126, 1965.
3. Report. Committee on Tissue Culture Viruses and Vaccines. Continuously cultured tissue cells and Viral Vaccines. Science 139:15-20, 1963.
4. Pirquet C.F., and Schick B. Serum Sickness. Translated by Bela Schick. Wiliams Wilkins, Baltimore, 1951.
5. Hull R.N., Minner J.R. and Mascoli G.C. New agents recovered from tissue cultures of monkey kidney cells. III. Recovery of additional agents both from cultures of monkey kidney tissues and directly from tissues and excreta. American Journal of Hygiene 68:31-44, 1959.
6. Stones P.B. Production and control of live oral poliovirus vaccine in WI-38 human diploid cells. Joint WHO/IABS Symposium on the standardization of cell substrates for the production of virus vaccines, Geneva, Dec. 1976. Develop. biol. Standard., 37: 251-253. S. Karger, Basel 1977.

7. Christofinis G.J. and Finter N.B. The preparation of Interferon from lymphoblastoid cell lines. Proceedings of Symposium. Preparation, Standardization and clinical use of Interferon. Zagreb, June 8-9, 1977.
8. Pritchett R., Pedersen M. and Kieff E. Complexity of EBV homologous DNA in continuous lymphoblastoid cell lines. Virology 74:227-231. 1976.
9. Nilsson K., Giovanella B.C., Stehlin F.S. and Klein J. Tumorgenicity of human hematropoietic cell lines in Athymic mice. Inst. of Cancer, 19:337-344, 1977.
10. Fantes K.H. Workshop on Clinical Trials with Interferon. National Institute of Allergy and Infectious Diseases and National Cancer Institute, Bethesda. March 21-23, 1978.
11. Hughes M., Thomson, R.O., Knight P., and Stephen J. The Immunopurification of Tetanus Toxoid. J. Applied Bacteriology 37:603-621, 1974.
12. Beale, A.J. D antigen content in poliovaccine as a measure of potency. Lancet 2:1166, 1961.
13. Salk J., Cohen H., Fillastre, et al. Killed poliovirus antigen titration in humans. Developments in Biological Standardization: Standardization and use of vaccines in the developing countries. Proceedings 15th International congress, April 1978. International Association of Biological Standardization (in press).
14. van Wezel A.L., van Steenis G., Hannick Ch.A. and Cohen H. Developments in Biological Standardization: Standardization and use of vaccines in the developing countries. Proceedings 15th International Congress, April 1978. International Association of Biological Standardization (in press).
15. Pereira H.G., Valentine R.C., and Russell W.C. Crystallization of an Adenovirus Protein (The Hexon). Nature, 219:946-947, 1968.
16. Mautner V. and Pereira H.G. Crystallization of a second adenovirus Protein (the Fibre). Nature, 230:456-457, 1971.
17. Haase A.T., Mautner V. and Pereira H.G. The immugenicity of adenovirus type 5 Structural proteins. Journal Immunology, 108: 483-485, 1972.

RECOMMENDATIONS FOR INACTIVATED HUMAN VACCINES PRODUCED ON HUMAN DIPLOID CELLS

Robert Netter

Laboratoire National de la Sante
25, boulevard Saint-Jacques
75014 Paris France

Human diploid cell lines (HDC) have now been used for the production of vaccines for 15 years and different international meetings have formulated requirements for the characterization and control of new cell strains. (Proc. Symp. on Characterization and uses of human diploid cell strain. Opatija, Yugoslavia 1963. -Minutes fifth and sixth Meeting Committee on cell cultures, Permanent section on microbial standardization 1968 and 1971).

The WI-38 cell line originally used is now being replaced for production by MRC-5 cells which fulfill the requirements (Jacobs, Jones, Baille 1970 - Characteristics of a human diploid cell designated MRC5- Nature 227, 168-170.

In the past there have been problems in the use of primary monkey kidney cells for vaccine production because of the new pathogenic extraneous viruses (B Virus, SV-40, Marburg); over the last 15 years, however no pathogenic agent has ever been found in human diploid cells authorized for vaccine production. Moreover no immunologic disorders have appeared in the vaccines. So far, due to the purity of human diploid cell lines and to the highly sophisticated controls applied to them it has been mainly live virus vaccines that have been produced on such cells, and WHO Technical Report Series, n° : (1977) 610 has included recommendations for rubella live virus vaccine produced on HDC. The scarcity of monkeys, the risk of contamination of humans from animals or primary tissue culture, and the fear of extraneous pathogens resistant to inactivation processes have led national authorities and producers to use HDC for rabies, rubella and poliomyelitis vaccine production and even smallpox, mumps and measles vaccine in Yugoslavia. Suggested recommendations were given in 1974 by

R. Netter and F. Perkins for inactivated rabies vaccine in WHO Techniques de Laboratoire sur la Rage-OMS 1974, pp. 354-365. Due to the experience gained since that time in the routine production of rabies inactivated vaccine, it seems that these requirements could be made less stringent by comparing with requirements for primary tissue culture inactivated vaccine without impairing safety or innocuity and applied to other vaccines.

TERMINOLOGY

Human diploid cell line: cells at an early passage, derived from a single human tissue fulfilling all the requirements of the 5th Meeting Committee on cell culture 1968 or other requirements recognized by the National Control Authority as being equal or more sensitive and which have not given rise to side effects during trials in not less than 10,000 humans of vaccines using virus grown in such cells.

Cell seed: a number of ampoules of cell suspension that have been derived from a single ampoule at an early passage and kept frozen in liquid nitrogen. Each ampoule is numbered and labelled with the date of its preparation and number of cell doublings represented. Special care is taken to prevent any contamination with other cells. Use of each ampoule for any purpose must be recorded.

Virus seed: stock of virus of uniform composition from which the number of passage is authorized by a National Control Authority for vaccine production.

Virus harvest: a virus harvest or pool of virus harvests derived from one ampoule of cells.

Bulk: a quantity of virus obtained by growing in the cells from one or more ampoules of cells, filtered, concentrated and inactivated by an approved process. The bulk must at some stage be contained in a single container.

Final product: final sealed vials of vaccine filled during the same operation and identical as regard the risk of contamination.

GENERAL MANUFACTURING REQUIREMENTS

The WHO requirements in the General Requirements for Manufacture and Testing of Biological Substances, Revised 1966 shall apply (WHO Technical Resport 1966, n° 323).

1. Room shall be decontaminated before use for cell propagation or vaccine production.
2. No cell and no virus other than used for vaccine production shall be introduced in the area.
3. The personnel involved in the production of a vaccine shall be shown to be immune to the virus involved.

CONTROL OF PRODUCTION

Control of Material

Virus strain and seed lot system. The virus strain used for vaccine production must be identified by historical records, and shall be shown to be pure and the genetic markers recorded. Immunogenicity in humans must be established and no more than a fived number of subcultures from this virus seed as determined by the licensing authority is authorized without further clinical assay to establish satisfactory safety and immunity. In the seed lot system operating, it is essential that the seed and the production lot must satisfy all tests prescribed but they are not always similar as for poliovirus living vaccine made on HDC for which the seed could be produced on monkey kidney cells.

Cell substrate for virus production. The cell seed must fulfill all requirements of the 5th Meeting of the Committee on Cell Culture 1968. The number of cell doublings permitted is a characteristic of the cell line and is dependant upon the lifespan and stability of karyotype as well as upon the absence of heterotransplantability. The tests are outlined below. MRC5 for example tolerates more split than WI-38.

Animal serum that may be used in the growth medium complies with the WHO requirements for tests for the sterility of Biological Substances (WHO Technical Report Series (1973) n° 530, 52) including tests for the presence of mycoplasma and bacteriophage by a method approved by the licensing authority. When the vaccine virus is grown in the cells, the maintenance medium shall not contain animal serum so that final product contains less than one part per million.

Control of Virus Seed

The virus seed used in the production of vaccine must fulfill all the tests applied to a live vaccine.

Control cells. 10% of the cells used for the production of seed are set aside by taking 10% of cells of each Roux bottle at the time of splitting, pooling them, seeding for the following tests and keeping a certain amount in liquid nitrogen.

Karyotype analysis. Chromosome monitoring shall be done on a representative sample from the pooled cells. Cells in metaphase are examined as follows: 300 cells are examined for polyploidy; an exact count is made on 100 cells and a complete karyotype analysis is made on at least one cell. The upper limit of abnormalities is 9% for breaks, 5% for polyploidy, 3% for structural abnormalities and 2% for hyper diploidy.

Test for heterotransplantability. Inject 10 mice with 2.10^6 cells by the subcutaneous route and 10 mice as a control group with 1.10^6 KB, HeLa or HEP II cells by the same route. Each mouse will receive 0.25ml of antithymocyte serum by intraperitoneal route (rabbit serum against mice thymocytes) at day -1, +1, +3, +6. At the end of 21 days, all animals in both groups shall be killed,any tumor at site of inoculation recorded and eventually collected for histopathological examination. At least 80% of animals must survive the entire observation period. None of the mice injected with HDC will show tumor and more than 80% of mice injected with tumorigenic cells will develop a progressive growing tumor. Test of equivalent sensitivity with other immunosuppresive drug such as corticosteroid might be used.

Identity of cells. The cells shall be identified by tests involving the HLA system using cytotoxicity, complement fixation, or by any other genetic marker of equal sensitivity.

Test on cell cultures for extraneous agents. The cells shall be examined for the presence of extraneous agents by the following test.

Hemadsorption test: 25% of the total control cells shall be incubated at the same temperature and under the same conditions as is made production cells until the last harvest of virus. This time shall not be less than 14 days. At least 80% of control cells shall remain healthy and the cells shall be normal showing no cytopathy effects due to viruses. The presence of hemadsorbing agents is detected with 4 species of red cells : human O, guinea pig, monkey and chicken by incubation for 30 minutes between 2-10°C, 30 minutes at ambient room temperature, and finally 30 minutes at 34-37°C for monkey erythrocytes. No hemadsorption should take place.

Tests on the supernatant fluids of control cells. At the time of each virus harvest from the production bottles, aliquot of supernatant fluids from the control cells is collected, pooled, stored at -70°C and tested as follows:

-Test on primary monkey kidney cell : 10 ml of pooled fluid is inoculated onto primary monkey kidney cell cultures; the concentration of the inoculum in the maintenance medium should not exceed 1/4; the morphology of the cells shall remain unchanged after 14 days and hemadsorption test under section Test on cell cultures for extraneous agents shall be negative. A subculture is made after 14 days and the cells stained after 14 days; the cells should not show any specific inclusion bodies.

-Test on primary rabbit kidney cells: 10 ml of the pooled fluid is inoculated onto primary rabbit kidney cells and the cultures are observed for 14 days. None of the cells shall show abnormalities.

-Test on human cells : 10 ml of the pooled fluid is inoculated onto human cells different from those used for production; a primary human amnion or another diploid strain is suitable. The cultures are observed for 14 days, less than 20% of cells are eliminated for accidental reason and no virus is demonstrated.

Test of the cells in animals for extraneous agents. 10 one day old mice from 2 different liters and 10 adults mice of 18-20 g shall be inoculated with 10^6 cells by the intramuscular route for each of them. 5 guinea pigs weighing 350 g to 450 g and 5 rabbits weighing 1,5 kg $\mp$ 10% shall be inoculated with .2.10^6 cells by the intramuscular route. The animals shall be observed for 30 days, 80% of the animals shall remain healthy and no death or disease attributable to the cells is observed.

Test of the cells in fertile eggs for extraneous agents. 10 embryonated eggs 9-11 days old shall be inoculated each with 10^6 cells by the allantoic route. After 3 days incubation at 34 to 36°C, the allantoic fluid is collected and shown to be free from hemagglutinating viruses using guinea pig and chicken red cells incubating for 30 minutes at 2-10°C and for 30 minutes at ambient temperature.

CONTROL OF VIRUS SUSPENSION BEFORE CLARIFICATION

The virus titer is measured by an appropriate method using tenfold dilutions.

Marker tests. At least one marker test approved by the licensing authority shall be included.

Tests for extraneous virus in cell cultures. The virus harvest is mixed with an adequate amount of high titer antiserum to neutralize the whole of the virus suspension; neither the antiserum nor the antigen used for its preparation shall be of human or simian origin. 10 ml of the neutralized virus pool is inoculated in each of the following cells : human diploid cells but of a different strain from those used for virus production; Cercopithecus or erythrocebus monkey kidney cells and primary rabbit kidney cells using the virus/serum mixture in the maintenance medium at a dilution of less than 1/4. If serum is used for the growth of cells or for neutralization, it should have been demonstrated that it does not contain inhibitors for human herpes virus. Control cells that remain uninoculated shall receive the same amount of antisera and same medium. After 14 days, 10 ml of the pooled supernatant will be subcultured into similar cells and incubated for a further 2 weeks. No cytopathogenic effect shall be observed and no hemadsorption shall be shown using guinea pig red cells for incubating for 30 minutes at 2-10°C as well as 30 minutes at ambient temperature.

Tests for extraneous virus in animals. Animals are inoculated with the neutralized virus harvest by the following tests.

-At least 20 adult mice weighing 15 to 20 g shall be inoculated intraperitoneally with 0.25 ml and intracerebrally with 0.03 ml. Animals are observed for 21 days. If any animal die the test may be repeated. If animals die in the second test of the same cause as in the first test and this is due to the virus harvest the material fails the test.

-At least 10 one day old mice or 20 suckling mice less than 7 days old shall be given 0.1 ml intraperitoneal route and 0.01 ml intracerebrally; the animals are observed 21 days (deaths that occur within the first 24 hrs are discounted).

-At least 5 rabbits weighing 1.500 kg are inoculated with 0.25 ml by the intracerebral route, 2 ml by the subcutaneous route and 10 times 0.1 ml by the intradermal route. Animals are observed for 21 days and no sign of sickness should appear.

-At least 5 guinea pigs weighing 350 g are inoculated with 1 ml by intraperitoneal route. Animals are observed for 6 weeks and necropsied examining them for tuberculosis.

Microbial, fungal and mycoplasmal sterility. The virus harvest shall comply with the WHO Requirements for the tests for sterility of biological substances (WHO Technical report 1973, n° 530, 52).

CONTROL OF VIRUS VACCINE

As the virus seed was produced on the same cells as the virus vaccine and as the cells used for production proceed from a frozen stock which was demonstrated free from extraneous agent, the risk is minimized as compared with poliomyelitis inactivated vaccine prepared on primary monkey kidney cells (WHO Technical report 1966, n° 323).

Control cells. Though for inactivated poliomyelitis vaccine produced on primary simian cells, it is not required to set aside control cells, the problem is not similar because with HDC there are more than 10 splits before seeding the virus. Animal serum used in growth medium is controlled as regard microbial sterility, mycoplasma and bacteriophages. 2.5% of the cells will be set aside as control cells before seeding the virus instead of 10% for virus seed.

Karyotype studies and or identity of cells. At least one of these test shall be performed as in sections entitled Karyotype analysis and Identity of cells instead of two tests for virus seeds.

TESTS IN CELL CULTURE FOR EXTRANEOUS AGENTS

Hemadsorption test. The test is carried on 25% of the control cells as in section Test on cell cultures for extraneous agents with only guinea pig red cells, instead of 4 different red cells species.

Inoculation of supernatant. The test is carried on human and simian cells only as in section Tests on the supernatant fluids of control cells instead of 3 tissue culture cells for virus seed.

VIRUS HARVEST BEFORE CLARIFICATION

Microbial and fungal sterility. Test is performed on 20 ml according to WHO Technical report 1973, n° 530, 52.

Absence of mycoplasma. Test is performed on 10 ml by an appropriate method such as US.21 CFR 610.12 April 1977.

Mycobacteria. Test is performed on 10 ml by an appropriate *in vitro* method such as described in the US 21 CR 630 16 April 1977.

Titration. After filtration or clarification, the harvest is titered to assure that the antigenic mass will be sufficient after inactivation. If attenuated virus strain is used for vaccine production, not less than one genetic marker has to be rechecked.

Virus harvest following concentration and inactivation. The virus harvest is concentrated by an appropriate method. Inactivation has to proceed immediately after filtration. Chemicals used should be demonstrated not to release potentially harmful derivatives in final product. The inactivation curve with at least 3 points is determined for the first five lots of vaccine rabies production or if any changes can interfere with the results. For poliovirus vaccine of course it should remain compulsory. The total inactivation time is at least three time, the minimum necessary to obtain viral inactivation.

Microbial and fungal sterility. Test is performed in 10 ml according to WHO Technical report 1973, n°, 530, 52.

Testing of inactivation and absence of extraneous virus. Standardization of the volume to be controlled is difficult as the risk for humans of contracting a serious disease is not comparable and volume of vaccine produced highly different (45 ml is recommended in WHO Technique on rabies and 500 ml in WHO Technical Report no° 323 for poliomyelitis).

Human diploid. Half of the retained inactivated suspension is inoculated the concentration of it in the homologous tissue culture medium not exceeding 1/4; the inactivating agent should have been previously excluded or neutralized. Viruses growing rapidly such as poliovirus are inoculated preferably on monolayers. Viruses growing slowly such as rabies virus may be inoculated in cell suspension and split twice at one week interval. Absence of infectivity is controlled by direct examination of the monolayer for poliomyelitis; for rabies vaccine, the monolayers are examined 7 days later by immunofluorescence (hemadsorption with guinea pig red cells at 4°C and ambient room temperature, and the supernatant collected 7 days after the last split is inoculated on 10 mice (12-15 g) by interacerebral route) (duration of observation 21 days).

Heterologous cells. Half of the retained inactivated suspensions is inoculated the concentration of it in the tissue culture medium not exceeding 1/4 and the inactivating agent being excluded. Selection of cells depends on the virus vaccine; for poliomyelitis primary monkey kidney cell monolayers are used; for rabies, sensitive baby hamster kidney cell line is used and absence of infectivity is controlled as described for HDC in section Human diploid.

Tests on the bulk material. For monovalent vaccine, final bulk is obtained by diluting the inactivated concentrated harvest or adding a stabilizing medium for freeze drying. For polyvalent vaccine, there is a mixing of each fraction in adequate proportion.

Microbial fungal and mycoplasmal stability. Test is performed according to WHO Technical Report 1973, n°, 530, 52.

Testing of absence of live poliomyelitis virus. The test is performed only for poliomyelitis vaccine on 1500 ml as described in section entitled Testing of inactivation and absence of extraneous virus.

Innocuity on monkeys. The test is performed only for poliomyelitis vaccine, as described in 5.3.1 of the WHO Technical Report 1966, n° 323, pp 34-35.

TESTS ON FINAL VACCINE

Identity and potency test. The potency test is described in the specific monographs. This test is not compulsory if stability test is performed as in section entitled Stability test.

Microbial and fungal sterility. Test is performed according to WHO Technical Report 1973, n° 530, 52.

Innocuity test. At least 5 mice weighing 17 to 22 g shall be inoculated with one human dose by the intraperitoneal route. Animals must remain healthy for one week. At least 2 guinea pigs weighing 250 to 350 g are inoculated with 5 human dose by the intraperitoneal route. Animal must remain healthy for one week.

Stability test. This test should be compulsory only for freeze dried vaccine. After incubation at 37°C, for varying length of time approved by the licensing authority depending on the virus, the vaccine should fulfill the potency test requirements described in the particular monographs.

Pyrogen test. After dilution 1/10 one human dose of vaccine per Kg of rabbit is inoculated by intravenous route. Total increase of the temperature of the 3 rabbits should not exceed 1.25°C.

THE SPECTOR OF MALIGNANCY AND CRITERIA FOR CELL LINES AS SUBSTRATES FOR VACCINES

Jonas Salk, M.D.

The Salk Institute for Biological Studies
P. O. Box 1809
San Diego, California 92112

While preparing for this meeting, I felt like Rip van Winkle awakening from a 20-year sleep to discover how much has happened and has been learned in the interim.

Those of us whose careers coincided with the introduction of a new generation of vaccines, produced from viruses grown in the chick embryo or in tissue culture, had far fewer technical possibilties available to us than exist today - but we also had far fewer limitations. Today, with the overwhelming increase in potential created by the explosion of knowledge in molecular and cellular biology, and in the technology of cell cultivation, the demanding standards for purity, safety, and effectiveness of biologicals are incomparably higher, but restrictions and limitations upon what one is allowed to do have also been greatly increased.

I recall the days when influenza virus vaccine was prepared by harvesting allantoic fluid admixed with chick embryo blood cells, from deliberately torn embryonic membranes, and then removing the blood cells with the adsorbed influenza virus to be concentrated by elution into physiological salt solution (1). Such crude preparations revealed the antibody-producing capacity and demonstrated the effectiveness of formalin-treated influenza virus in reducing the incidence of disease (2).

It was approximately 26 years ago that the first human subjects inoculated with experimental poliovirus vaccines received formalin-treated fluids from cultures of tissue fragments of monkey testis or kidney embedded in a chicken plasma clot, nourished with chick embryo extract and ox serum ultrafiltrate (3).

In appreciating the present while reflecting upon the past, I began to feel more like Alice in Wonderland than like Rip van Winkle. The possibilities that now exist make it seem that we will soon be able to do almost anything with cultured cells in the production of biologicals for human use. However, there are rate-limiting factors. Those of the past were different from those of today; and whatever they are now, will be still different in the years to come. In the past, such factors were related primarily to limitations in knowledge and technology; some have been related to fear evoking beliefs, such as the "theoretical" possibility of transmitting a neoplasia-inducing factor.

Such fears are not altogether new. In the early 1950's experimental vaccines against influenza and poliomyelitis admixed with incomplete Freund's adjuvant were studied in human subjects. (3,4). When a poliovirus vaccine so prepared was proposed for field testing, questions were raised about the theoretical possibility that the mineral oil in the adjuvant might cause cancer which would not become evident until 20 years later. Accordingly, this approach was abandoned for poliomyelitis. By this time, however, influenza vaccine emulsified in incomplete Freund's adjuvant had already been injected into human subjects, and follow-up studies made 10 and 18 years later revealed that the incidence of cancer was the same in the control and the test groups (5,6). Approximately 26 years have now elapsed and, if desired, another follow-up study could be carried out.

I cite this experience because of its analogy to the question of the "theoretical" risk of cancer induced by vaccine made from virus grown in continuously propagating cell lines. In the mid 1950's such cells were proposed and studied for their usefulness in the production of poliovirus vaccine. It was hoped that such substrate would be generally useful in preparing vaccines for human use. The resistance then encountered still exists today.

At the Fourth International Poliomyelitis Conference held in Geneva in 1957, Dr. Syverton expressed the opinion "that the specter of malignancy hovers over the question as to whether or not one can use an independently (continuously) propagating cell..." and urged "that investigations be done, rather than resorting to hypotheses and speculation." (7). My own view is reflected in the question I posed, "Is this Conference going to end without some firm and definitive idea that before another Conference... some definite steps will be taken to supplant the monkey kidney in the preparation of poliovirus vaccine and thus to eliminate one of the tremendous problems confronting those interested in vaccine technology?" (7).

At that time we had prepared killed poliovirus vaccines from a continuously propagating cell line then believed to have been of monkey origin (8) and since shown to have characteristics indistinguishable from those of HeLa cells (9,10). In an effort to determine its tumorigenic potential, we observed, in monkeys receiving relatively large inocula, that the capacity of the cell to be tolerated and to induce local tumor formation depended upon the degree of immunocompetence of the host; nonmetastasizing tumors were observed at the inoculation site in tuberculous monkeys in the tuberculin skin test-negative stage of their disease (11). Ruptured cells, cell extracts, or poliovirus vaccines made from such cells were not tunorigenic and large numbers of viable cells inoculated intravascularly failed to induce tumor formation.

Continued resistance to the use of continuously propagating cells for the preparation of vaccine is evident in the proceedings of the 1967 Conference on Cell Cultures for Virus Vaccine Production: "Until clear evidence to the contrary is available, I believe that we must consider...any marker of abnormality in cultured cells as a possible indicator of in vivo tumorigenicity. But the weight to be given to (this belief)...must obviously be balanced against the value of the vaccine....if we use diploid cell lines of human origin... one doubt would still remain - perhaps even unaltered diploid human cells, intact or extracted, could become tumorigenic if inoculated into man... Since this remote possibility cannot be tested by inoculating man, the use of diploid cell lines of animal origin could provide an additional safeguard (12). It is inconceivable to me that a system will be found which is completely free of any potential risk; so the real problem is one of minimizing the risk by applying all of the monitoring and assay techniques which we now have or may develop in the future... It will also place a heavy burden of responsibility on the control authority which has to approve the use of such vaccines, and as Dr. Enders has so aptly said '...to determine if the game is worth the candle.'...Development of fractionated, nucleic acid-free, viral antigens would appear to be the bright hope of the future and the needed technology is now being developed... Perhaps 5 years from now we will look back and remember this period as the end of the 'horse-and-buggy' days of vaccine. (13)."

Moore (14) expressed still another view: "...cancer cells grown under ordinary tissue culture conditions are probably less dangerous in a vaccine than the unseen microorganism, either bacterial or viral, that they may carry."

The same issues are again being considered at this symposium. However, the context has changed in that we have acquired further knowledge and experience and the need for extending the range of cell substrates for vaccine production has been recognized. The interplay between belief and knowledge, as rate-controlling factors, in the evolution of cell substrates for vaccine production, is indicated in Figure 1. Primary monkey kidney cells were the first cells to be

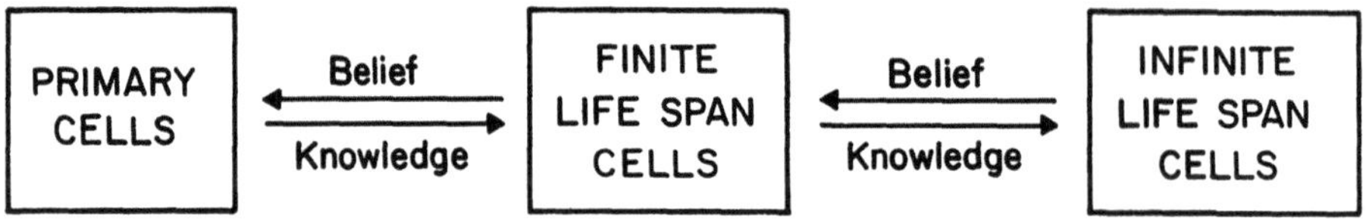

Figure 1. Rate-limiting or accelerating factors in the evolution of cell substrates for vaccine production.

introduced and used in the preparation of poliovirus vaccine; after a period of resistance, human diploid cell lines of finite lifespan were accepted and are now in use. The question before us concerns the use of infinite lifespan cell lines. The factor that limits their acceptance is the belief in the possibility that vaccines made from viruses grown in such substrates might be oncogenic.

The reasons to consider a broader range of cell substrates include the need to improve the quality and supply of present vaccines, and to develop new vaccines. An insufficient supply of monkeys, or of fetal bovine serum for maintaining diploid cell lines on a large scale, as well as the restricted range of virus susceptibility of presently approved cell lines, limit the rate of improving the quality and supply of current vaccines and of developing new vaccines. The proposed solution is to include the possibility of using infinite lifespan cell lines. However, as shown in Figure 2, karyology and lifespan have been employed as criteria for exclusion of cell lines to avoid uncertainty about the risk of oncogenicity. The black symbols indicate the reasons for exclusion; cells with characteristics which have been avoided are those shared by malignant cells, i.e. finite lifespan cells which are euploid and nontumorigenic are acceptable.

Figure 3, redrawn and revised from Petricciani (15),indicates that if aneuploidy and infinite lifespan are removed, the only criterion for exclusion would be tumorigenicity. If this is not justified, then we are being deprived of the benefit of a useful substrate of practical value in preparing vaccines for human use on a large scale and with greater economy. It would seem important, therefore, to determine whether or not tumorigenic cells lines are associated with an oncogenic risk when vaccines for human use are prepared in them. If there is no demonstrable risk, if there is a clear advantage to their use, then all infinite lifespan cell lines could be placed in competition for their efficiency and economy in producing the desired vaccines. If cells which are tumorigenic, as compared with those which are nontumorigenic, are superior by a sufficiently wide margin, then an effort should be made to determine whether or not viruses prepared in such substrates, or extracts of the cells themselves, are oncogenic (Figure 4). It would be of

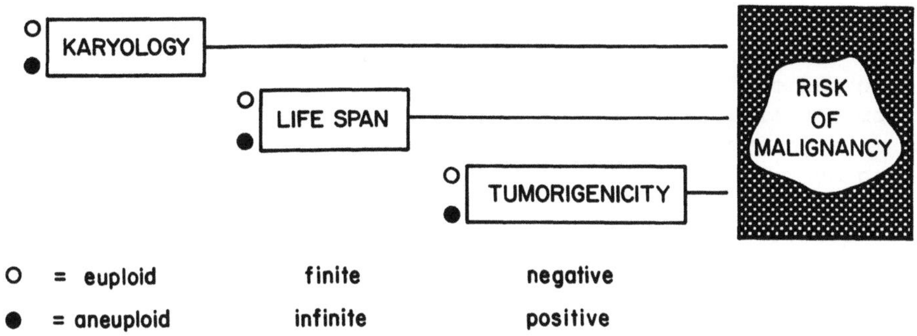

Figure 2. Margin of safety in characteristics of cell lines

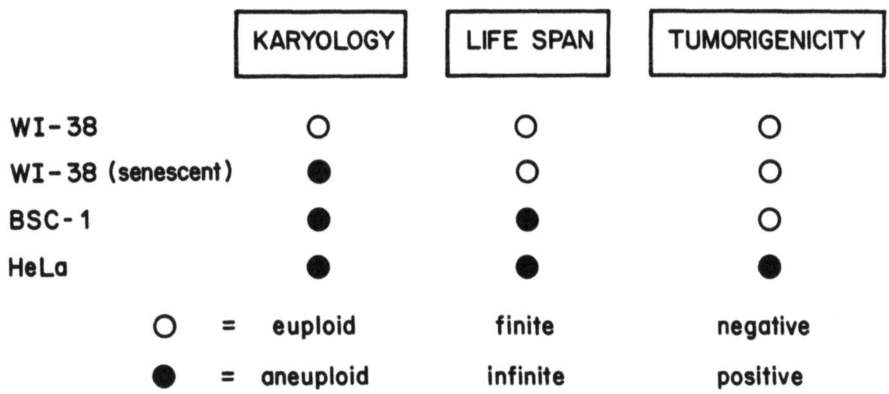

Figure 3. Characteristics of cell lines which might be considered for use in biologics production (adapted from J. Petricciani, 1978)

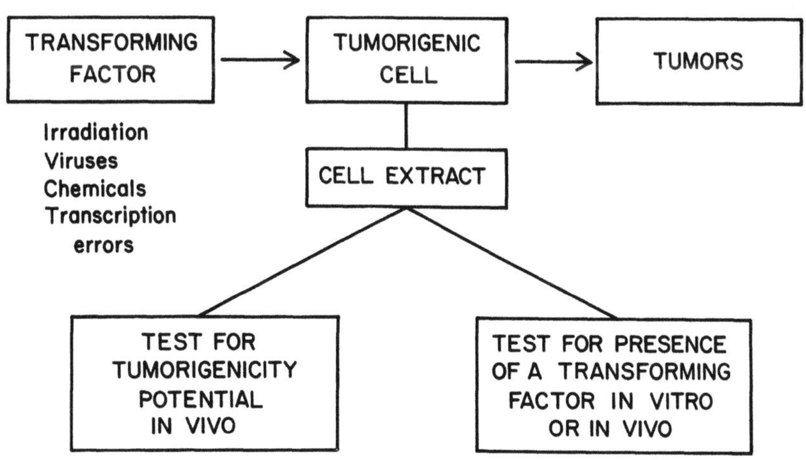

Figure 4. Malignant potential of extract of tumorigenic cells (A Testable Question)

value to carry out a systematic program to determine, in general, whether or not homologous tumorigenic cells contain transmissible oncogenic factors either in extracts of such cells or in vaccines made from noninfectious viruses propagated in them. Such studies can be made in appropriate mouse and rat systems with observations extending over the lifespan of these animals.

The question of criteria for inclusion or exclusion of cell lines as substrates for preparing virus for vaccines for human use should be open for further consideration until sufficient data can be produced to determine the presence or absence of oncogenic potential of vaccine produced from viruses propagated in such cells. The practical and societal advantages of including infinite lifespan cell lines for making vaccines for human use, provide incentive for a systematic exploration to settle this question.

REFERENCES

1. Francis, T. Jr., Salk, J. A simplified procedure for the concentration and purification of influenza virus. Science 96: 499-500. 1942.
2. Commission on Influenza: A clinical evaluation of vaccination against influenca. J. Am. Med. Assoc. 124:982-985. 1944.
3. Salk, J.E., Bennett, B.L., Lewis, L.J., et al. Studies in human subjects on active immunization against poliomyelitis. 1. A preliminary report of experiments in progress. Ibid. 151:1081-1098. 1953.
4. Salk, J.E., Contakos, M., Laurent, A.M.,et al. Use of adjuvants in studies on influenza immunization. 3. Degree of persistence of antibody in human subjects two years after vaccination. Ibid. 151:1169-1175. 1953.
5. Beebe, G.W., Simon, A.H., Vivona, S. Follow-up study on army personnel who received adjuvant influenza virus vaccine. Am. J. Med. Sci. 247:385-406. 1964.
6. Beebe, F.W., Simon, A.H., Vivona, S. Long-term mortality follow-up of army recruits who received adjuvant influenza virus vaccine in 1951-1953. Am. J. Epidemiol. 95:337-346. 1972.
7. Salk, J. (Discussion Section). Poliomyelitis, Fourth International Poliomyelitis Conference. Edited for the International Poliomyelitis Congress. Philadelphia, J.B. Lippincott, 1958. pp. 289-290.
8. Salk, J., Ward, E.N. Some characteristics of a continuously propagating cell derived from monkey heart tissue. Science 126:1138-1139. 1957.
9. Clausen, J.J. Syverton, J.T. Comparative chromosomal study of 31 cultured mammalian cell lines. J. Nat. Cancer Inst. 28: 117-145. 1962.
10. Greene, A.E., Coriell, L.L.,Charney, J.A. Rapid cytotoxic antibody test to determine species of cell cultures. Ibid. 32:779-796. 1964.

11. Salk, J. Immunological paradoxes - theoretical considerations in the rejection or retention of grafts, tumors, and normal tissue. Ann. N.Y. Acad. Sci. 162:365-380. 1969.
12. Deinhardt, F. Viral oncogenesis and results of inoculation of normal and transformed cells into nonhuman primates. Cell Cultures for Virus Vaccine Production. NCI Monograph 29, Washington, DC, US Government Printing office. 1968. pp. 327-331.
13. Stevenson, R.E. Viral oncogenesis and results of inoculation of normal and transformed cells into nonhuman primates - discussion. Ibid, pp. 33-334.
14. Moore, A.E. Evidence for oncogenic properties of heteroploid cell lines - discussion. Ibid, pp 291-292.
15. Petricciani, J. (paper in this volume).

CELL SUBSTRATES FOR BIOLOGICALS OTHER THAN VACCINES

Introductory Remarks

Jan Vilček, M.D.

Department of Microbiology, New York University
School of Medicine, 550 First Avenue, New York,
NY, 10016

Pioneers of contemporary cell culture methods have long believed that animal or plant cells in culture could yield many useful substances, including hormones, immunoglobulins, enzymes and other biologically active agents. However, progress in the actual use of cell cultures for the large-scale production of biologically active substances has been slower than some of us had anticipated. Until recently the only pharmaceutical products that required the use of cell cultures were some viral vaccines. This situation is changing.

Two cell culture products with therapeutic activity in man are discussed in this session: interferon and urokinase. Interferons are very much in the limelight these days. Long known as natural glycoproteins able to protect cultured cells as well as the animal organism from various virus infections, interferons are also endowed with many other biological activities. Foremost among these is the inhibitory activity on the growth of some tumors that may be based on the direct inhibition of cell multiplication as well as on the modulation of immune responses by interferon. The demand for human interferon to be used in clinical trials in viral infections and in cancer requires a fresh look at the available cell substrates for interferon production.

Although applications of urokinase in clinical medicine are of a very different nature, some problems connected with its large scale production and safety testing are common to urokinase and interferon. Technological advances made in the course of development of urokinase production in human embryonic cells in culture should be useful for the development of other products from cultures of anchorage-dependent cells.

While the era of the production of useful biologically active substances in tissue culture has barely gotten underway, we already must start looking for other alternatives. Will recombinant DNA technology provide practical means for the economical production of animal cell constituents in prokaryotic cells? Indications are that this is a very realistic alternative, although many technical problems remain to be solved.

For those of us who may be troubled by the inability to relate problems of everyday science to broader philosophical issues of our times, Dr. Robert R. Wagner finishes off this session with a beautifully written essay "Molecular Biology and Public Health." At first I was very much intrigued by the title of Dr. Wagner's paper. After readying his essay I was delighted. You will be too.

INTERFERON AS A CELL PRODUCT

Jan Vilcek, M.D.

Department of Microbiology, New York University
School of Medicine, 550 First Avenue, New York,
NY 10016

INTRODUCTION

Everyone is aware of the long lag period between the general notion of the clinical potential of interferon, developed some twenty years ago, and the actual application of interferon in clinical practice. Today we are at a stage when practical sources of human interferon for clinical trials are finally being developed and it is becoming possible to produce quantities of interferon that - though still relatively modest - will allow a thorough evaluation of interferon's clinical potential.

Not only has progress on the road towards developing methods of interferon production been slow, there were several unexpected turns to be taken along the way. Originally we were modestly hoping to develop the production of interferon for use as an antiviral agent. But we have learned that interferons play a much broader role: they have immunoregulatory activities (1,2), they affect various cellular functions including cell division (3,4), and possibly they may be of some value as anticancer drugs (5). On the other hand, there is evidence suggesting that in some virus infections, rather than protecting the host, interferon may be a factor contributing to the pathology of the disease (6,7).

DISTINCT SPECIES OF HUMAN INTERFERON

We started out by trying to develop the production of a human interferon. In the process we have learned to our great surprise that there are at least three distinct species of human interferon (Table 1). Le interferon is the major interferon species produced in cells of lymphoid origin, whereas F interferon is the predominant species obtained after induction of nonlymphoid cells. Furthermore, there is a third species variously referred to as "immune" or "type II" interferon that is preferentially produced in cultures of lymphocytes, particularly in cultures enriched for T cells. Although it is not clear whether T lymphocytes are the only source of this interferon, we refer to it as T interferon, as originally suggested by Wietzerbin et al (8).

These three interferons have many distinct properties(9); in fact, they must have very different primary structures and they are most likely products of distinct structural genes. They all fit the definition of interferon in that they are induced proteins (probably glycoproteins) that on incubation with competent cells produce a characteristic antiviral state. But they can be distinguished from each other on the basis of many unique properties (Table 1).

Monospecific antisera are available against Le and F interferons. These sera will neither neutralize nor bind any of the two heterologous interferons (10,11). There is no antiserum yet available against T interferon, but this interferon has no affinity for either anti-Le or anti-F serum. Until specific antibody for this third type of interferon becomes available, T interferon is best defined as being antigenically non-Le, non-F. We cannot be sure that T interferon is indeed one homogeneous species. Incidentally, not every interferon that is in the literature described as "immune" or "type II" interferon belongs in this category. When we recently analyzed some mitogen-induced or antigen-stimulated lymphocyte interferons, we found that these preparations contained the garden variety Le interferon, or, sometimes, a mixture of Le and T interferons. Thus, it is important to be aware of the fact that what some people have referred to as "immune" or "type II" interferon may or may not be the real McCoy.

In addition to distinct antigenic characteristics, the three interferons differ in their physicochemical properties. Since it was shown that the activity of many interferons can be recovered after treatment with the anionic detergent sodium dodecylsulfate (SDS), molecular weights of interferon can be accurately and reproducibly measured (12,13). Le interferon is composed of two molecular weight forms: the lower one (approximately 15,000 daltons) is most likely an unglycosylated or underglycosylated form of the higher molecular weight form (about 20,000 daltons).

There are a few recent reports suggesting that T interferon is more efficient in its nonantiviral actions than the other interferons. Thus T interferon might, according to preliminary data, be more efficient in its immunoregulatory, cell growth inhibitory and

Table 1. Some distinguishing properties of the three known species of human interferon

Interferon species	Approx. mol.wt.	Stability at pH 2	Relative efficiency of antiviral action in human fibroblasts	Antiviral activity in bovine or porcine cells
F	22,000	Stable	Very high	Very low
Le	20,000 + 15,000	Stable	High	High
T	?	Labile	Low	Absent or very low

tumor growth inhibitory activities than Le and F interferons. It has been postulated that the primary function of T interferon may be immunoregulation or some other cellular regulatory activity, rather than antiviral action (1,14). However, very little definitive information is available about T interferon, mainly because good production and purification methods have not yet been worked out. In this respect, the study of T interferon is at a stage that is reminiscent of the level of understanding (or the lack of understanding) we had about the other species of interferons 10 or 15 years ago. It seems safe to offer the prediction (for whatever it's worth) that at one of the future W. Alton Jones Cell Science Center Symposia we are likely to discuss in more detail the large scale production and possible clinical applications of this most interesting albeit somewhat elusive entity - T interferon.

SOME PROPERTIES OF Le and F INTERFERONS RELEVANT TO THEIR CLINICAL USE

We shall from now on discuss only human Le and F interferons. Some investigators may not yet be totally convinced that these two interferons are distinct entities. In fact, the question is often asked whether it is not possible that Le and F interferons are identical gene products that are processed in different ways by the respective producing cells or that differ merely in the amount or type of carbohydrate present. We did very seriously consider these possibilities but most available data tend to refute them. Some of the experimental evidence is summarized in Table 1.

Le and F interferons differ in some characteristics of their antiviral action. In a recent study from our laboratory (15) human skin fibroblasts in culture were exposed to 10, 30 or 100 units/ml of either F or Le interferon for varying times. The cultures were then washed to remove extracellular interferon and inoculated with a high dose of vesicular stomatitis virus (VSV). To assess the degree of resistance attained, the inhibition of VSV yield was measured. Although the antiviral states attained after 6 hours of contact with comparable doses of either Le or F interferon were similar, the rate of development of antiviral state was faster with F interferon than with Le interferon. In other words, after a short contact of cells with interferon, i.e., after 5, 15, or 45 minutes, the inhibition of VSV yield was greater in cells incubated with F interferon than in cells incubated with comparable doses of Le interferon.

Thus, it appears that in cultures of human fibroblasts F interferon is more efficient in inducing the antiviral state. This finding raises the question whether this difference between F and Le interferons is a general characteristic that will be seen in any type of cell or whether it is perhaps the result of a certain tissue specificity of action of the two human interferons - one being preferentially produced, and also more active, in nonlymphoid cells and the other having a predilection for cells of lymphoid origin. We don't know the full answer but at least one piece of experimental evidence suggests that the two interferons do show a certain tissue specificity in their action: Einhorn and Strander (16) tested the cell growth inhibitory action of Le and F interferon on two types of tumor cells in culture. In their hands F inhibited more efficiently osteosarcoma cells and Le interferon was more strongly inhibitory against lymphoblasts. Recently Kuwata in Japan compared the cell growth inhibitory action of Le and F in transformed cells of nonlymphoid origin and concluded that F was more efficient (17).

There are still too few examples of such possible tissue specificity and more comparative studies of the actions of Le and F interferons in different cells are obviously needed. However, taken together, these data lead to the logical conclusion that we should not promote the production of one interferon species at the expense of the other, but that both Le and F interferon production systems should be developed alongside. To argue that we need one interferon but not the other would be no more logical than to propose that one broad spectrum antibiotic is all we need to treat bacterial infections or that a vaccine containing poliovirus type 1 is all that is needed to protect against all types of poliovirus.

Differences between Le and F interferons extend to the realm of pharmacokinetic properties. Because of the earlier and much greater availability, the bulk of clinical data was obtained with preparations of interferon prepared in buffy coat cell (leukocyte) cultures containing Le interferon as the predominant species. This interferon is usually administered intramuscularly. It was shown that much of the intramuscularly injected Le interferon eventually enters into circulation and can be recovered from the blood in biologically active form (18, 19). In initial clinical trials with preparations of interferon derived from diploid fibroblasts, intramuscular administration apparently failed to produce the expected levels of activity in the circulation (20). These findings are of some concern because, obviously, interferon cannot be expected to show systemic activity if it does not reach the site at which it is supposed to exert its action. This problem could be due to one of the following events: (1) F interferon is either bound or inactivated at the injection site and does not enter into the circulation at a sufficiently high level; (2) F interferon enters the circulation but is rapidly inactivated in the blood; (3) F interferon enters the circulation but is very rapidly cleared from the blood.

These different possibilities can be tested experimentally and we undertook to do the necessary experiments. Because fibroblast interferon is still quite precious, we chose an animal that is smaller than man and, therefore, requires less interferon - the rabbit.

It was reported that various body fluids, including human serum, exert an inactivating effect on interferon and that this inactivating effect is more pronounced with interferon preparations derived from human fibroblasts than with buffy coat (leukocyte) interferon (21,22). We were unable to find a marked inactivation of either fibroblast or leukocyte interferon by whole rabbit blood at 37°C. No measurable inactivation occurred after 1 hour of incubation with either whole blood or with control tissue culture medium. (A slight increase in interferon activity was observed after 1 hour of incubation with rabbit blood - probably due to the relative inaccuracy of the interferon assay rather than to resurrection of interferon activity or some other supernatural event.) After 12 hours of incubation a slight drop in interferon activity was demonstrable both in the presence of blood and in the control mixture. Since the half-life of interferon in the circulation is measured in minutes rather than hours, direct inactivation of interferon by some component of the blood is unlikely to be of pharmacokinetic significance.

We have also tested the other possible explanations for the lack of appropriate blood levels after intramuscular injection of fibroblast interferon. Recent evidence from our laboratory shows that after intravenous injection in rabbits both Le and F interferons are cleared at similar rates. In view of this result the remaining possible explanation for the difference in the pharmacokinetic properties of fibroblast and leukocyte interferon could be either greater retention or greater inactivation of fibroblast interferon at the muscular site of injection. A direct comparison of blood levels produced in rabbits injected intramuscularly with comparable doses of human leukocyte or fibroblast interferon showed the following: 1) the maximum levels of interferon activity attained in the blood were

2 to 4 times higher in rabbits injected with leukocyte interferon 2) peak levels of interferon activity were attained somewhat sooner with leukocyte interferon than with fibroblast interferon 3) titers of leukocyte interferon in the blood decreased faster, so that between 5 and 24 hours after intramuscular administration interferon titers tended to be higher in animals injected with fibroblast interferon (J. Vilček, I.T. Sulea and Y.K. Yip, in preparation).

Taken together these results suggest that F interferon is probably not irreversibly inactivated at the injection site but that, in comparison with Le interferon, its resorption is retarded. This difference should not preclude clinical trials with fibroblast interferon in man but, until some means are found to improve the resorption from the muscular site of injection, intravenous administration appears to be the preferred route of injection of this interferon in man.

BRIEF REVIEW OF CELL CULTURE SYSTEMS AVAILABLE FOR LARGE-SCALE PRODUCTION OF HUMAN INTERFERON

Since this symposium is concerned with cell substrates, it is not entirely inappropriate to discuss the main features and relative merits of the available cell culture sources of human interferon. Three systems have been thoroughly characterized as to the optimal conditions leading to the production of maximal interferon yields and the properties of interferons produced (Table 2).

The first system employs cultures of primary buffy coat cells derived from fresh human blood, stimulated to produce interferon by the addition of Sendai or Newcastle disease viruses (NDV). Lymphocytes are probably the major source of interferon. This system has been perfected over more than a decade of continuous use, mainly as a result of sustained efforts of K. Cantell and his colleagues in Finland (23,24). The overwhelming majority of clinical trials has utilized interferon produced from this source. Provided that a sufficient supply of fresh human blood can be secured, the methodologies of interferon production and purification now available are relatively simple (25). The major obstacle for industrial scale production is the large volume of human blood needed. Buffy coat cells from one unit of blood yield on the average about 2 x 10^6 interferon reference units (24). Since 25-50% of activity is lost during the purification process, buffy coats from about 4 or 5 units of blood are needed for the production of interferon sufficient for a single average daily dose for one patient. However, since after collecting buffy coats the remaining blood can be used for transfusion or other purposes, one should not underestimate the potential of this time honored and thoroughly tested cell substrate.

Table 2. Human interferon species produced in tissue culture systems available for large-scale production

Cell Culture	Inducer	Approx. proportion produced %Le	%F
Primary buffy coat cells	NDV or Sendai virus	99	1
Lymphoblastoid cell lines (Namalva)	NDV or Sendai virus	85	15
Foreskin fibroblast cell strains (FS-4)	Poly(I)·poly(C) "superinduction"	<0.1	>99.9

Suspension cultures of serially propagated lymphoblastoid cells are another widely employed source of human interferon. Out of a large number of lymphoblastoid cell lines scrutinized (all derived from patients with Burkitt's lymphoma), the Namalva line has proved to be the best interferon producer (26). The optimal conditions for interferon production have been examined in several recent studies (27,28). The important appeal of this cell substrate lies in its relatively easy adaptability to advanced large-scale technology akin to that employed for products derived from suspension cultures of microorganisms. The virtues of this cell substrate have been sufficiently enunciated by other speakers at this Symposium (see contributions by M. Hilleman and J. Beale). Possible biosafety problems (or the lack of them) associated with products from permanent cell lines derived from tumors have also been broadly discussed by other contributors.

It is noteworthy that Le is the predominant interferon species produced in both primary buffy coat cultures and cultures of lymphoblastoid cells after stimulation with NDV or Sendai virus (Table 2). However, in preparations from Namalva cells about 15% of total activity is associated with F interferon whereas buffy coat cell derived material contains only traces of F interferon activity (29). In contrast, preparations obtained from the third available source-diploid fibroblasts, usually do not contain any detectable Le component. Optimal interferon yields from fibroblast cultures are obtained after stimulation with the synthetic double-stranded RNA, poly(I)·poly(C), in conjunction with judiciously timed treatments with inhibitors of cellular protein and RNA synthesis (30,31). This regimen is referred to as superinduction (Table 3). Lower interferon yields are obtained when cultured fibroblasts are stimulated with viruses, but it is interesting that virus infection of fibroblasts results in the synthesis of small amounts of Le interferon (11). Thus, the relative proportion of Le and F interferons produced depends not only

Table 3. Outline of interferon superinduction schedule used for foreskin fibroblast strains

Hour	Procedure
0	Cultures are washed once with buffered saline; induction medium is added, consisting of poly(I)·poly(C) (5-20 μg/ml) and cycloheximide (10-50 μg/ml) in serum-free MEM
5	Actinomycin D (final conc. 1-5 μg/ml) is added in a small volume of MEM directly to the induction medium on the cultures
6	Induction medium is removed and discarded; the cultures are washed 4 times with saline; production medium is added, consisting of MEM with 0.1% (final conc.) human plasma protein fraction, U.S.P. or 0.1% human serum albumin
30	Production medium containing crude interferon is collected

on the type of producing cell but also on the nature of the inducer, albeit the latter is apparently of lesser import in determining what species of interferon is synthesized.

The technology of interferon production in anchorage dependent human fibroblasts is, of course, very different from that employed with the other two systems. In order to produce interferon in fibroblast cultures on an industrial scale, existing methods used for the production of virus vaccines in diploid fibroblasts would have to be scaled up and new technological improvements introduced. This route is being taken by some groups.

It appears that cell strains derived from neonatal foreskins are the most readily available source of this interferon. Interferon yields obtainable from foreskin cell strains are generally higher than from the existing licensed strains, WI-38 and MRC-5. Interferon yields also vary somewhat among cell strains derived from individual foreskins (32). A high producer foreskin strain, FS-4, originally established in our laboratory, has been widely employed for experimental interferon production in dozens of laboratories throught the world. Several new foreskin cell strains producing interferon yields similar to those obtained from FS-4 cells have been recently identified among strains derived at the W. Alton Jones Cell Science Center (33). One of these cell strains, FS-35, is being made available for wide use in interferon production by the Antiviral Substances Program of the National Institute of Allergy and Infectious Diseases.

Unlike some of the other cell substrates discussed at this Symposium, FS-4, FS-35 and other foreskin cell strains are diploid, they have a limited life span and they have other characteristics of normal untransformed cells. In these respects they are comparable to human diploid cell strains with an established safety record in the production of viral vaccines. Interferon preparations with similar characteristics can be produced in cultures of tumor cells, for instance in a line derived from human osteosarcoma (34). However, since interferon yields from this transformed cell line are not higher than from suitable diploid cell strains, there seems to be no important incentive to use it for large-scale interferon production. This attitude might change in the future if new lines or clones of transformed cells are found that produce higher yields of F interferon than diploid cell strains. The isolation of SV40 transformed cell lines producing large quantities of F interferon has been described - alas without complete documentation (35).

So, for now, diploid cell strains remain the most widely used cell substrate for the production of "fibroblast" interferon. In view of what appears to be a rather general mood for the relaxation of rules governing the use of cells in the production of biologicals, I feel a trifle embarrassed to advocate the use of such an uninspiring, safe cell substrate. To paraphrase Dr. Hilleman's parlance, we are left with no dragons to kill. Or are we?

ACKNOWLEDGEMENTS

Original work from the author's laboratory was supported in part by Contract NO1-AI-02169, Grants AI-07057 and AI-12948 from the National Institute of Allergy and Infectious Diseases, and by the German Federal Ministry for Research and Technology in conjunction with Rentschler Arzneimittel GmbH & Co.

REFERENCES

1. Johnson, H.M., Baron, S. Interferon: Effects on the immune response and the mechanism of activation of the cellular response. CRC Critical Reviews in Biochemistry 4: 203-227, 1976.
2. De Maeyer, E., De Maeyer-Guignard, J. Effect of interferon on cell-mediated immunity. Tex. Rep. Biol. Med. 35: 370-374, 1977.
3. Paucker, K., Cantell, K., Henle, W. Quantitative studies on viral interference in suspended L cells. III. Effect of interfering viruses and interferon on the growth rate of cells. Virology 17: 324-334, 1962.
4. Gresser, I., Brouty-Boyé, D, Thomas, M-T, et al. Interferon and cell division. I. Inhibition of the multiplication of mouse leukemia L 1210 cells *in vitro* by interferon preparations. Proc. Nat. Acad. Sci. 66: 1052-1058, 1970.

5. Strander, H. Anti-tumour effects of interferon and its possible use as an antineoplastic agent in man. Tex. Rep. Biol. Med. 35: 429-435. 1977.
6. Rivière, Y., Gresser, I., Guillon, J-T., et al. Inhibition by anti-interferon serum of lymphocytic choriomeningitis virus disease in suckling mice. Proc. Nat. Acad. Sci. USA 74:2135-2139. 1977.
7. Ida, S., Hooks, J.J., Siraganian, R.P., et al. Enhancement of IgE-mediated histamine release from human basophils by viruses: Rose of interferon. J. Exp. Med. 145: 892-906. 1977.
8. Wietzerbin, J., Stefanos, S., Falcoff, R., et al. Immune interferon induced by phytohemagglutinin in nude mouse spleen cells. Infect. Immun. 21: 966-972. 1978.
9. Vilček, J., Havell, E.A., Yamazaki, S. Antigenic, physicochemical and biological characterization of human interferons. Ann. N.Y. Acad. Sci. 284: 703-710. 1977.
10. Havell, E.A., Berman, B., Ogburn, C.A., et al. Two antigenically distinct species of human interferon. Proc. Nat. Acad. Sci. USA.72: 2185-2187. 1975.
11. Havell,E.A., Hayes, T.G., Vilček, J. Synthesis of two distinct interferons by human fibroblasts. Virology 89: 330-334. 1978.
12. Mogensen, K.E., Cantell, K. Stability of human leukocyte interferon towards heat. Acta Path. Microbiol. Scand. B. 81: 382-383. 1973.
13. Stewart, W. E. II, De Clercq, E., De Somer, P. Stabilisation of interferons by defensive reversible denaturation. Nature 249: 460-461. 1974.
14. Sonnenfeld, G., Mandel A.D., Merigan, T.C. Immunomodulating activitiy and characterization of type II interferon. Fed. Proc. 36: 1229. 1977.
15. Gardner, L.J., Vilcek, J. Initial interaction of human fibroblast and leukocyte interferons with FS-4 fibroblasts. J. Gen. Virol. (in press).
16. Einhorn, S., Strander, H. Is interferon tissue specific? - Effect of human leukocyte and fibroblast interferons on the the growth of lymphoblastoid and osteosarcoma cell lines. J. Gen. Virol. 35: 573-577. 1977.
17. Kuwata, T., Fuse, A., Suzuki, et al. Comparison of suppression of cell and virus growth in human transformed cells by leukocyte and fibroblast interferon. J. Gen. Virol. (in press).
18. Cantell, K., Pyhälä, L., Strander, H. Circulating human interferon after intramuscular injection into animals and man. J. Gen. Virol. 22: 453-455. 1974.
19. Jordan, G.W., Fried, R.P., Merigan, T.C. Administration of human leukocyte interferon in herpes zoster. I. Safety, circulating antiviral activity and host responses to infection. J. Inf. Dis. 130: 56-62. 1974.
20. Weimar, W., Heitjink, R.A., Schalm, S.W. et al. Fibroblast interferon in HB_SAg-positive chronic active hepatitis. Lancet ii: 1282. 1977.

21. Harmon, M., Greenberg, S., Couch, R. Effect of human nasal secretions on the antiviral activity of human fibroblast and leukocyte interferon. Proc. Soc. Exp. Biol. Med. 152: 598-602. 1976.

22. Cesario, T.C., Tilles., J.G. Inactivation of human interferon by body fluids. Tex. Rep. Biol. Med. 35: 443-448. 1977.

23. Strander, H., Cantell, K. Production of interferon by human leukocytes in vitro. Ann. Med. Exp. Biol. Fenn. 44: 265-273. 1966.

24. Cantell, K., Hirvonen, S., Mogensen, K.E. et al. Human leukocyte interferon: Production, purification, stability, and animal experiments. The Production and Use of Interferon for the Treatment and Prevention of Human Virus Infections - In vitro. Edited by C. Waymouth. Monograph no. 3, pp/ 35-38. 1974.

25. Cantell, K. Hirvonen, S. Large-scale production of human leukocyte interferon containing 10^8 units per ml. J. Gen. Virol. 39: 541-543. 1978.

26. Strander, H., Mogensen, K.E., Cantel, K. Production of human lymphoblastoid interferon. J. Clin. Microbiol. 1: 116-117. 1975.

27. Zoon, K.C., Buckler, C.E. Bridgen, P.J. et al. Production of human lymphoblastoid interferon by Namalva cells. J. Clin. Microbiol. 7: 44-51. 1978.

28. De Ley, M., Billiau, A., De Somer, P. A semi-continuous system for the production of human interferon in lymphoblastoid cell cultures. J. Gen. Virol. 40: 455-458. 1978.

29. Havell,E.A., Yip, Y.K., Vilček, J. Characteristics of human lymphoblastoid (Namalva) interferon. J. Gen. Virol. 38: 51-59. 1977.

30. Billiau, A., Joniau, M., De Somer, P. Mass production of human interferon in diploid cells stimulated by poly I:C. J. Gen. Virol. 19: 1-8. 1973.

31. Havell,E.A., Vilček, J. Production of high-titered interferon in cultured of human diploid cells. Antimicrob. Ag. Chemother. 2: 476-484. 1972.

32. Mozes, L.W., Havell, E.A., Gradoville,M.L.et al. Increased interferon in human cells irradiated with ultraviolet light. Infec. Immun. 10: 1189-1191. 1974.

33. Vilček, J., Havell, E.A., Gradoville, M.L. et al. Selection of new human foreskin fibroblast cell strains for interferon production. Human Interferon, Production and Clinical Use. Edited by W. R. Stinebring, P.J. Chapple. New York, Plenum Press, pp. 101-118. 1978.

34. Billiau, A., Edy, V.G., Heremans, H. Human interferon: Mass production in a newly established cell line, MG-63. Antimicrob. Ag. Chemother. 12: 11-15. 1977.

35. Berthold, W., Tan, C., Tan, T.H. Purification and in vitro labeling of interferon from a human fibroblastoid cell line. J. Biol. Chem 253: 5206-5212. 1978.

UROKINASE AS A CELL PRODUCT

Joseph C. Fratantoni, M.D.

Bureau of Biologics

Bethesda, MD

UROKINASE AS A CELL PRODUCT

Fibrin deposition occurs in health and disease. The formation of fibrin results from the sequential activation of proteases according to a scheme now well known as the coagulation cascade. A variety of stimuli can initiate this cascade and the resultant fibrin may be involved in physiological (healing) or pathological (vascular obstruction) sequelae. Resolution of fibrin deposits is accomplished by the action of an enzyme system which has as a substrate the insoluble fibrin polymers. This is the fibrinolytic system and, akin to the coagulation system, consists of sequential activation of proenzymes. Plasminogen, a circulatory β-globulin, is converted to plasmin (fibrinolysin) by tissue or plasma kinases (activators). One of these activators, urokinase, is produced by renal cells and is excreted in the urine. The teleologically inclined will want to consider that this system maintains the urinary tract free from obstructing fibrin clots. Urokinase has been found therapeutically useful, thus stimulating a search for production sources other than human urine. One successful approach is based on cell culture, thus the inclusion of this topic in your program. I should now like to consider with you the therapeutic use of urokinase, the methods of isolation, and some recent data regarding its function.

Therapy with Urokinase

The discovery of plasminogen activators led to attempted use in vascular occlusion and the development of thrombolytic therapy. The agent initially employed was streptokinase, a plasminogen activator derived from hemolytic streptococci. Clinical studies demonstrated that it could induce a fibrinolytic state which resulted in dissolution of experimental thrombi (1). Considerable data accumulated, dealing with venous thromboembolism, myocardial infarction and other thromboembolic disorders. Some problems with the streptokinase available at that time stimulated development of preparative procedures for urokinase.

The Committee on Thrombolytic Agents (CTA) of the then National Heart Institute established standards for urokinase and two manufacturers (Abbott Laboratories and Sterling-Winthrop) provided large quantities of the material prepared from fresh urine.

Two major clinical trials were then undertaken: The Urokinase Pulmonary Embolism Trial (UPET) (2) and the Urokinase-Streptokinase Pulmonary Embolism Trial (3) (USPET). Both Trials measured the responses of patients with angiographically proven pulmonary embolism to randomly assigned treatment. UPET studied a 12 hour infusion of urokinase vs. heparin, while USPET compared urokinase for 12 hours vs. urokinase for 24 hours vs. streptokinase for 24 hours. The end points included physical examination and an array of diagnostic methods then available. In brief, the trials demonstrated that these agents would result in more rapid dissolution of an embolus than observed with heparine alone. However, there was no difference in mortality between the various groups (4). Table I summarizes the data from these two major trials.

Additional therapeutic trials have been conducted utilizing patients with myocardial infarction trials showed no effect of thrombolytic therapy when death was the endpoint. Venographic study of deep vein thrombosis demonstrated that thrombolytic therapy would result in more rapid clearing of affected vessels than heparin. Streptokinase has recently been licensed for use in patients with deep vein thrombosis or with pulmonary embolism. Urokinase is approved for use in patients with pulmonary embolism.

Production of Urokinase

The presence of a plasminogen activator in human urine was first described in 1951 (5) and was given the name "urokinase" in 1952 (6). Processing on a commercial scale was available by the early 1960's. A 12 hour course of therapy, for use in UPET or USPET, required collection and processing of 1500 liters of urine which yielded a 4 million CTA unit dose of urokinase. At the time, the cost of such a dose was $1200.

Concern regarding the logistics of urine collection and processing on a massive scale led to the development, by Abbott Laboratories, of a commercial process for production of urokinase from cell cultures. This was made possible by a series of accomplishments beginning with that of Barnett & Barone in 1959 (7). They

TABLE I

Summary of Results Obtained in UPET & USPET Trials

	UPET		USPET		
	Heparin	12 hr. UK	12 hr. UK	24hr. UK	24 hr. SK
Number of patients	78	82	59	54	54
Change in angiographic severity score	0.54	1.78	1.66	1.76	1.70
Decrease in % lung perfusion	8.3	24.1	20	29.2	18.5
Change in pulmonary artery pressure (mmHg)	-1.1	-6.2	-7.3	-7.5	-5.3
Deaths	7	6	4	5	5

demonstrated that plasminogen activator (and other protease activators) were produced by cultured human epithelial carcinoma cells and by primary monkey kidney cells. Subsequent work established presence of a plasminogen activator in established cultures of canine renal cells.

In 1969, Bernik & Kwaan published their observations which demonstrated that fibrinolytic activity was present in cultures of human kidney cells, that the greatest yields came from fetal kidneys and that the fibrinolytic agent was immunologically identical to urokinase isolated from human urine (8).

The details of the Abbott production process are not available for public disclosure, but the procedure can be described in outline. Human fetal kidney cells are established in culture using a medium containing serum. When the cells reach confluency the growth medium is removed, residual serum washed off and protein-free production medium added. The urokinase is harvested from the medium using ion exchange and gel filtration techniques. Quality control measures include fibrinolytic assay, pyrogenicity and thromboplastin contamination.

Multiple Molecular Forms of Urokinase

It has been known from some years that urokinase can be found in multiple molecular sizes (9). There appear to be two major forms: low molecular weight (LMW, 33,000 daltons) and high molecular weight (HMW, 57,000 daltons). Of particular interest was the observation that urinary urokinase contains predominantly the HMW form and tissue culture urokinase contains predominantly the LMW form. It appears that the HMW form is the native structure and that it is converted to LMW by cellular proteases.

It was of great practical importance to compare the efficacy of these two forms. As noted above clinical trials were conducted using urinary, therefore HMW, urokinase. LMW urokinase results in identical fibrinolytic effects and, as demonstrated by careful, though limited, clinical studies, also gives similar therapeutic effects.

A seeming paradox arose when it was noted that HMW was more active when assayed with standard enzyme kinetic methods. Dr. G. Murano, from the Bureau of Biologics, has looked into this question and the data I will now refer to were collected in his laboratory and are presented with his permission (10). The greater activity of HMW is observed only when measurements are made using plasminogen concentrations at plasma levels. In systems containing very high substrate (plasminogen) concentrations, both forms are quivalent. This led to the consideration of an inhibitor in plasma and recent experiments have shown that when urokinase is incubated in plasma: 1. If residual urokinase activity is plotted vs. time, HMW demonstrates a more rapid decay curve. 2. This decay is related, at least in part, to the presence of antithrombin III in plasma. 3. Both HMW and LMW form irreversible complexes (apparently stoichiometric) with antithrombin III.

Heparin, which exerts its anticoagulant activity via antithrombin III, accelerates the urokinase decay rates, but not dramatically. Thus, the more active HMW form is also inactivated more rapidly in plasma, accounting for the equivalent hemostatic and clinical results. These studies will continue and may shed light on the fundamental and practical aspects of thrombolytic therapy.

REFERENCES

1. Brogden, R.N., Speight, T.M., Avery, G.S. Streptokinase: a review of its clinical pharmacology, mechanism of action and clinical uses. Drugs 5: 357-445, 1973.
2. A National Cooperative Study: The Urokinase Pulmonary Embolism Trial Circulation 47: Suppl. 2: 1-108, 1973
3. A National Coopertive Study: The Urokinase Streptokinase Pulmonary Embolism Trial. JAMA 229: 1606-1613, 1974.
4. Fratantoni, J.C., Ness, P., Simon, T.L. Thrombolytic Therapy. N. Eng. J. Med. 293: 1073-1078, 1975.
5. Williams, J.R.B. The fibrinolytic activity of urine. Brit. J. Exp. Pathol. 32: 530-537, 1951.
6. Sobel, G.W., Mohled, S.R., Jones N.W., et al. Urokinase: an activator of plasma profibrinolysin extracted from urine. Am. J. Physiol. 171: 768, 1952.
7. Barnett, E.V., Barone, S. An activator of plasminogen produced by cell culture. Proc. Soc. Exp. Biol. Med. 102: 308-311, 1959.
8. Bernik, M.B., Kwaan, H.C. Plasminogen activator activity in cultures of human tissues. J. Clin. Invest. 48: 1740-1746, 1969.
9. Lesuk, A., Terminiello, L., Traver, J.H. Biochemical and Biophysical Studies of Human Urokinase. Thrombosis et Diathesis Haemorrhagica 18: 293-294, 1967.
10. Murano, G., et al., Inhibition of high and low molecular weight urokinase by plasma inhibitors. Circulation 58:II-136, 1978.

PRODUCTION OF BIOLOGICALS BY PLANT CELL CULTURES

Donald K. Dougall, PhD

W. Alton Jones Cell Science Center

Lake Placid, NY 12946

INTRODUCTION

The inclusion of a chapter on the production of biologicals by plant cell cultures in this volume is both appropriate and timely. The topic is timely because the production of a biological by plant cell cultures on commercial scale may be operational within five years. The primary purpose of this chapter is to review the current state of knowledge of plant cell culture as it bears on the production of biologicals.

Many different types of compounds isolated from plants have pharmacological activity. The range of chemical types and pharmacological activities have been listed by Farnsworth and Bingel (16) and Tyler et al (62). A list of some of the known active compounds from plants is shown in Table 1. In addition to those with pharmacological activity, man utilizes fragrances, flavors, colorings and gums as well as enzymes obtained from plants.

Potential Problems with Whole Plants as Sources of Biologicals

Potential problems associated with the use of whole plants as sources of biologicals can be summarized as environmental, economic and political. Yields can be affected by environmental factors such as excessively cold seasons, drought or loss of crops due to natural disasters. Economic difficulties can arise from variation in yield of desired compound and changing costs of growing or harvesting plants. As more land is converted to food production to meet population needs, less land will be available for non-food i.e. drug crop usage thus raising the cost of land for production.

Table 1

Examples of Biologicals Obtained from Plants

Pure Compounds of Medical Importance	
a) Drugs	Morphine, codeine, quinidine, pilocarpine, d-tubcurarine, reserpine
b) Cardiac Stimulants	Digoxin
c) Antitumor Compounds	Vinblastine, vincristine, maytansine, bruceantin, taxol, homoharringtonine
d) Steroids	Diosgenin, Solasodine
Extracts or Crude Botanicals	Ipecac, Cascara, opium
Insecticides	Pyrethrins, rotenone

Labor costs continue to rise. For those plants which are harvested from wild populations, continued use may exceed the replenishment rate of the natural populations leading to decreased availability and increased price. Changes in currency exchange rates may also affect the availability of raw material. For plants from foreign countries, political problems such as warfare, internal strife or discriminatory governmental action may alter the supply of source plants. These aspects together with the difficulty of adapting some native plants to new habits or even to cultivation, suggests that alternative sources need to be found. In some cases chemical synthesis of the biological or of an alternative synthetic compound, may be an economic alternative. However, in some cases the molecule is sufficiently complex to make synthesis uneconomic and therefore a biological source must be used.

PLANT CELL CULTURES AS A SOURCE OF BIOLOGICALS

Plant cell cultures may be an alternative to whole plants as sources of biologicals. This idea was expressed in the patent issued to Routienn and Nickell in 1956 (51). They proposed that cells of a species producing a desired biological would be grown in fermentors as are bacteria or fungi, harvested and the desired biological recovered. Plant cell cultures has advanced dramatically since that time so that a concept which was then a theoretical possibility is now moving into the realm of practical reality. In terms of the production of biologicals by plant cell cultures, the period 1956 to present divides into two eras with 1973 being the year separating them. In the period 1956 - 1973 low yields of biologicals were obtained from plant cell cultures. However, since

1973 the yields obtained from cell cultures have begun to approach and in a number of cases, exceed those obtained from whole plant.

Yields of Biologicals from Plant Cell Cultures

The productivity of plant cell cultures is usually assessed by comparing the yield as a percentage of dry weight, of the biological from cell cultures with that obtained from whole or parts of plants of the producing species. This comparison is widely used and convenient. However, it is superficial. The yield of biologicals in plants is strongly altered by the conditions of growth and the genotype of the plant (64). Accumulation of biologicals in specific parts of plants e.g. seeds or roots, leads to different yields depending on which part of the plant is analyzed. The comparison of yield in cell cultures with that from the plant must be made bearing these variables or factors in mind. This is particularly important when the yields are approaching parity and is much less important at the extreme situations.

Yields of Biologicals in Plant Cell Cultures: Pre-1973 Era

The yields obtained in early measurements of biologicals in plant cell cultures were lower than those found in the plant, generally by several orders of magnitude. This work has been reviewed by Butcher (6), Staba (53), Puhan and Martin (47) Krikorian and Steward (39), and Constable et al (8). It is important to note, however, that the biologicals characteristic of the plant species from which the cultures were initiated, were in fact produced in the cultures. This finding indicates that the capacity of the cells to produce, or the genetic information needed to produce the biologicals was not completely lost from the cultures.

Because in many instances plants could be regenerated from the cultures, it was possible to verify that the capacity for synthesis of specific biologicals had not been lost from the cell cultures in a number of cases. These studies were performed with cell cultures of *Coptis japonica* (28), of 11 members of the family *Papaveraceae* (27), of *Scopolia parviflora* (57), of *Nicotiana tabacum* (59), of *Digitalis purpurea* (24) and of *Datura innoxia* (23). In all cases the cell cultures produced low levels of alkaloids. The alkaloids present in the cell cultures were also qualitatively different from those in the species of origin. The plants regenerated from the cell cultures recovered the patterns and levels of alkaloids characteristic of the species. Thus in these cases the capacity for synthesis of specific compounds is retained during culture.

In addition to the studies on production of biologicals by plant cell cultures during this period some other aspects of plant cell culture were undergoing major development. These include the improvement of media for the growth of plant cells (19), the establishment of cell cultures from a very wide range of plant species and the development of methods for cloning plant cell cultures (5,43). While these developments were not in response to the needs associated with the production of biologicals, they did in fact significantly influence the events of the period after 1973.

Yields of Biologicals in Plant Cell Cultures: Post-1973 Era

While it is true to say that, as a rule, prior to 1973 the yields of biologicals in plant cell cultures were very low relative to that from the whole plant there were in fact a few exceptions to the rule. In particular, Kaul et al (36) obtained good yields of the steroid diosgenin from cell cultures of *Dioscorea deltoidea* in 1969. However, since 1973, there has been a substantial increase in the number of reports showing yields of biologicals approaching those found in the plant. These reports are shown in Table 2 together with the species, compounds isolated, and yields. In the cases of tissue cultures from *Catharanthus roseus*, *Morinda citrifolia*, *Symphytum officinale*, *Coleus blumei*, *Daucus carota* and *Nicotiana tabacum* (producing ubiquinone) the yields are from five to several hundredfold higher than recorded for the whole plant or its parts. It is to be noted that in the cases of the tissue cultures of *Ruta graveolens* and *Andrographis paniculata*, a series of novel compounds were identified and characterized. These compounds were not found in the whole plant.

Taken together the cases listed in Table 2 suggest that plant cell cultures which give yields of biological approaching those found in the plant should be achievable for any plant species.

COMPONENTS OF THE CULTURE SYSTEM WHICH INFLUENCE YIELDS OF BIOLOGICALS

While the number of examples given in Table 2 suggests that high yields of biologicals in plant cell cultures can be achieved, the information in Table 2 does not show how to achieve cell cultures giving high yields of biologicals. As a starting point in the determination of how to achieve high yielding cell cultures, one can inquire into the differences in practices or knowledge of plant cell cultures before and after 1973 which have led to the increased yields. The answers to such an inquiry would very likely identify the factors responsible for the increase in yields. The same end can be achieved by asking which components of a culture system influence the yields of biologicals. This wider question allows additional information to be brought to bear on the problem of achieving satisfactory yields of biologicals in plant cell cultures. Such an inquiry leads to the identification of three major components of culture systems which are important. Each of these will be discussed separately.

TABLE II

Compounds Isolated from Plant Tissue Cultures in Yields Approaching or Exceeding those Found in Whole Plant Listed by Year of Report

Year	Species	Compound	Yield from tissue culture	Yield from plant	Reference
1969	*Dioscorea deltoidea*	Diosgenin	26 mg/g dry wt		36
1971	*Ruta graveolens*	Coumarins	isopimpinellin 1.2-1.6 rutamarin 0.1-0.2 rutacultin 0.4-0.6 mg/g dry wt	not present traces only not detected	54
	Ruta graveolens	Alkaloids	kokusaginine edulinine	not present not present	54
	Andrographis paniculata	Paniculides,A, B,C (sesquiterpene lactones)	paniculides A,B,C	not found in plants	7
1974	*Symphytum officinale*	Glutamine	924 μmoles/g dry wt	leaf 3.5 μmoles/g dry wt	60
	Panax ginseng	Ginseng saponins	0.38% fresh wt	0.3-3.3% fresh wt	30
	Panax quinquefolium		0.41% fresh wt		

TABLE II
(continued)

Year	Species	Compound	Yield from tissue culture	Yield from plant	Reference
1975	*Morinda citrifolia*	Anthraquinones	900 μmoles/g dry wt	root 110 μmoles/g dry wt	70
	Cassia tora	Anthraquinones	0.334% fresh wt	seeds 0.209% dry wt	58
	Nicotiania tabacum	Ubiquinone	0.5% mg/g dry wt	leaf circa 16 μg/g dry wt	25,26
	Daucus carota	Cyanidin	10% dry wt	trace	2,11
	Scopolia japonica	Proteinase inhibitors	41.0 mg/g dry wt	stem 12.5 mg/g dry wt leaf 24.1 mg/g dry wt root 37.1 mg/g dry wt	41
1976				buds bark root	65
	Acer pseudoplatanus	Phenolics	132 mg/g dry wt	72 32 14 mg/g dry wt	
		Flavanols	68 mg/g dry wt	38 15 7 mg/g dry wt	
		Leuco-anthocyandin	305 OD units/g dry wt	45 12 8 OD units/g dry wt	
	Stephania cepharantha	Biscoclaurine alkaloids	10-22.9 mg dry wt total alkaloids	tuber 8.2 mg/g dry wt aerial part 1.0 mg/g dry wt	1
	Nicotiana rustica	Nicotine	0.291% dry wt	root 0.25% leaf 0.35% of dry wt	55
	Catharanthus roseus	Serpentine	0.5% dry wt		10

TABLE II
(continued)

Year	Species	Compounds	Yield from tissue culture	Yield from plant	Reference
	Ephedra foliata	Pseudophedrine	2.25% dry wt	stem 0.6% dry wt	37
1977	Catharanthus roseus	Serpentine Ajmalacine	1.3% dry wt	0.26% dry wt	72
	Trigonella oculta	Diosgenin Gitogenin Tigogenin	0.37% dry wt 0.14% dry wt 0.05% dry wt		29
	Coleus blumei	Rosmarinic Acid	11% dry wt	2.1% dry wt in stems 4.3% dry wt in leaves	48,71
1978	Galium mollugo	Anthraquinones	12.9 mg/g dry wt		4,66
	N. tabacum	Nicotine	3.4% dry wt		45
	Lithosperum erythrorhizon	1,4-Naphtoquinones	0.57% dry wt	root 1.16% dry wt	42

Genotype of the Plant used to Initiate Cell Cultures

The genetic control of alkaloid production and accumulation in *Lupinus*, *Nicotiana*, *Datura* and *Papaver* species is discussed by Waller and Nowacki (64). It is clear that the yield of alkaloids in plants is under genetic control. The genetic information in the cells of the plant will set limits on the capacity of the cells to produce the biological of interest. Evidence to support this was obtained by Zenk et al. (72) and Roller (50) whose data showed that on the average, high yielding plants of *Catharanthus roseus* gave cultures whose yield of the alkaloid, serpentine was higher than that in cultures from low yielding plants. However, the data of both Zenk et al. (72) and Roller (50) have examples where this general observation does not hold. Variation between cultures derived from different plants was observed not only in yield of alkaloids but also in the growth of the cultures (72). Ohta et al. (46) have observed wide differences in both growth and nicotine production in cultures derived from individual plants of *Nicotiana tabacum* var. Bright Yellow. It is reasonable to assume that both yield of biologicals and growth of cells in cell culture are influenced by the specific genetic constitution of the plant from which the culture was initiated. However, the nature of the genetic information which influences growth in cell culture is not known. It is not possible to predict which genotypes will give high yields of biologicals and/or good growth of cultures. Thus one of the requirements for achieving high yielding cultures is to ensure that a wide range of genotypes are examined to find those individuals which give high yields of biologicals and suitable cell culture behavior in the same culture.

Nutritional and Cultural Conditions

Cultures of plant tissues can be initiated in nutrient media containing a range of formulations of minerals, vitamins, and sugar (19). The major difference between the nutrient media used to initiate cultures of different species lies in the concentrations and nature of the growth stimulating compounds of the classes, auxin and cytokinin. Because auxins and cytokinins are the major factors in the initiation of cultures and because these compounds influence the yields of specific chemicals, they will be discussed first.

Auxins and Cytokinins. The presence of an auxin at a suitable concentration in the medium is required for the initiation of tissue cultures from primary explants. The presence of a cytokinin at a suitable concentration may also be required. In addition to their effects on initiation and maintenance of growth in tissue cultures, auxins and cytokinins have substantial effects on yields of biologicals. This will be illustrated by considering the effects of the three widely used auxins: indoleacetic acid (IAA), naphthalene-acetic acid (NAA) and 2, 4-dichlorophenoxyacetic acid (2,4-D). In carrot cell cultures growing on medium with auxin at 10^{-5}M, the production of cyanidin was very low with IAA and NAA but with 2,4-D was stimulated 10-fold (2). With suspension cultures of Morinda citrifolia,the production of anthraquinones was increased 30-fold by substitution of NAA for 2,4-D at 2 ml/l (70). Furuya et al. (18) showed that the production of nicotine by tobacco cells in culture occurred with IAA at 1 mg/l but did not occur with 2,4-D at 1 mg/l. The production of naphthoquinones by cultures of Lithospermum erythrorhizon occurred in the presence of IAA but not 2,4-D (56). These examples serve to show that the use of different auxins leads to widely different amounts of biologicals produced by a plant cell culture. They also show that in different cultures, different auxins give maximum yields. Consequently, depending on the auxin used cell cultures may or may not produce a desired biological despite the fact that the cultures are capable of producing the biological.

Other medium components and culture conditions. The yield of biologicals in plant cell cultures is influenced by many components of the medium in addition to auxins and cytokinins. It is also influenced by culture conditions. These effects have been discussed by Dougall (12). One can find at least one case where each component of the culture medium and each aspect of culture conditions, have effects on yield. Thus it is not possible to draw general conclusions about the effects of specific components of culture media or any aspect of culture conditions, on yield. However it is clear that a medium and a set of culture conditions for maximum yield should be established when a new culture system is being developed.

Cultures Contain Cells which give Different Yields of Biologicals

Evidence that cultures contain cells giving different yields of biologicals has been given by several authors. Tabata and Hiraoka (55) have demonstrated that clones derived from cultures of Nicotiana rustica differ in their yield of nicotine. Clones from a culture of Nicotiana tabacum differed over a 30-fold range in nicotine yield (45). Clones from Catharanthus roseus cultures gave a range of yields of 0-1.4% dry weight for serpentine and 0-0.8% dry wt for ajmalicine (72). The range of anthocyanin production in clones from wild carrot was 0.005-0.35 mg cyanidin/mg extracted dry wt (11). The range of solasonine yields from clones of a culture of Solanum laciniatum was 0-3.2% of dry wt (69). Clones with stable high yields have been described by Tabata and Hiraoka (55), Ogino et al. (45) and Zenk et al (72).

This heterogeneity in plant cell cultures may arise in the course of maintenance of cultures by serial passage. It may also arise from heterogeneity of cells in the primary explant. The sources and mechanisms of its generation are not known. The existence or development of heterogeneity in cell cultures may be part of the explanation for the often observed decline in yields when plant cell cultures are maintained by serial passage. If the culture conditions select for low yielding cell lines from a heterogeneous population, then selection for low yielding lines would continue during serial passage of the culture and lead to decline in yield of the culture.

Two general methods of cloning plant cell cultures have been described. The first involves the isolation and nurturing of single cells in microcultures. The most widely used microculture method is that of Muir et al. (43) in which single cells are grown, separated from a relatively large piece of cultured tissue by filter paper. Here the larger piece of tissue modifies the environment of the single cells and allows it to grow. The second general method is to localize single cells by plating them in a suitable medium containing agar and allowing them to grow (5). The conditions required to achieve plating efficiencies greater than 70% with tobacco cell cultures were established by Gibbs and Dougall (20). Kao and Michayluk (32) grew cells from Vicia hajastana from inocula containing 5 cells/ml. In all these cases, the suspensions of cells used were obtained by growing tissue in suspension cultures after which single cells and clumps containing limited numbers of cells were obtained by filtration of the suspensions. Recently Schulte and Zenk (52) have described a method for replicate plating of plant cells using a nylon screen to transfer a replica from one plate to another.

The selection of high yielding clonal lines from cultures of plant cells has been performed by examining the yields of each clone which is obtained and maintaining only those which give the highest yield. This approach has been termed "brute force." The time and effort required by this approach has been decreased somewhat in two cases (45, 72). The possibilities for the use of specific media to select for high yielding clones seems limited because it appears that the production of the special biological by the cell is not necessarily a requirement for cell growth. Despite these limitations on and difficulties with selecting high yielding clones, one can clearly improve the yield of biologicals by clonal selection.

There are, therefore, three components of culture systems which have significant effects on yields of biologicals. In attempts to develop high yielding cell cultures, the strategy used should identify and apply to the system, those aspects of each component which increase yields.

STATE OF PLANT CELL CULTURE TECHNOLOGY

The information used to identify components of culture systems which alter yield has been derived from laboratory experiments with cultures of small volume.

If plant cell cultures are to be grown in fermentors as part of a commercial process, then some technological requirements must be satisfied. These requirements are the ability to 1) grow plant cells in fermentors, 2) protect cultures from destruction 3) project costs of the process. The current status of these requirements is such that available information shows that they can be achieved.

Growth of Plant Cell Cultures in Fermentors

Growth of plant cells in batch cultures of greater than one liter have been performed in vessels with flat bladed turbine impellers (8,33,34,35,38,40,47) or with air agitation(33,41,61). Continuous and semi-continuous cultures of plant cells have been achieved (17,35,38,61,73). There are no great difficulties with simply growing plant cells under these conditions. Noguchi (44) described growth and productivity of tobacco suspension cultures in a two stage continuous-culture system where the vessels for each stage had a capacity of 35 liters, and in a batch cultures of 15,000 liters. Kato et al. (33) using tobacco cells showed that biomass yield was limited by oxygen if the initial K_{La} was below 15 hrs^{-1}. Evidence has been obtained which shows that in turbine fermentors, some plant cell cultures are destroyed by the shear stresses generated (74). Oxygen availability (74) and medium pH (3,75) also effect the production of biologicals by plant cells grown in fermentors.

Protection of Cell Cultures with High Yields

Methods to protect valuable plant cell cultures are in an early stage of development. There is now a growing body of evidence that plant cell cultures can be protected by storage in liquid nitrogen. Protection of cultures has been reviewed by Dougall (13,14) and cryogenic storage of plant cell cultures has been reviewed by Withers (67) and Withers and Street (68).

Costs

Only one description of a systematic effort to estimate the costs of production of biologicals using plant cell cultures has been given. Goldstein et al. (21) examined four cases at 3 production levels 10^4, 10^5, and 10^6 Kg/year. The first case was at the current state of technology of plant cell cultures and the remainder at a series of theoretical improvements on the current technology. For their model, they calculated the manufacturing costs to be $551/Kg and $17.25/Kg at the two extreme situations. At the same extremes, they calculated that selling prices $1,608 per Kg and $43 per Kg would allow profits of 15.7% and 9.1% of sales respectively after paying all costs and recovering the invested capital in 5 years.

COMMERCIAL USE OF PLANT CELL CULTURES FOR THE PRODUCTION OF BIOLOGICALS

So far in this paper, the term commercial has been used in the rather limited sense of a product which is produced in reasonable quantities. Commercial can also be used in the sense of an item which is sold on the open market without restriction on the size of the annual gross amount. In this wider definition of commercial, plant cell cultures are already being used as sources of biologicals on a commercial scale. Bethesda Research Laboratories Inc., (USA) is selling an acid pyrophosphatase isolated from tobacco cell cultures which is used as a reagent in studies of RNA structure and function. The use of this enzyme for removal of "caps" from mRNA was described by Efstratiadis et al (15).

Commercial use of plant cell cultures for production on a scale greater than that of a speciality item such as the acid pyrophosphatase mentioned above, appears not to have been achieved. There are two specific possibilities for development as commercial processes apparent from published material. The first of these is the production of ajmalicine which is used in the treatment of circulatory diseases. Ajmalicine is isolated from plants and is also prepared by reduction of the closely related alkaloid, serpentine. The production of serpentine and ajmalicine by cell cultures of C. roseus has been studied extensively (9,50,69,72). The available information on the C. roseus system shows that suitable cultures have been developed and that it is technically feasible to produce serpentine and/or ajmalicine on any scale desired. The question now is primarily one of economics. The second possible system is the conversion of β-methyl digitoxin into the cardiac stimulatory compound β-methyl digoxin by cell cultures from Digitalis lanata. Here conversion or biotransformation of an already complex molecule is to be achieved rather than the biosynthesis of the entire compound. In this case, a compound of low value (digitoxin) would be converted into one of high value (digoxin or a closely related active compound). Digitoxin is accumulated as a result of purification of digoxin from the mixture of cardiac glycosides extracted from D. lanata plants. The studies on the biotransformation of β-methyl digitoxin into its 12-hydroxylated derivative, β-methyl digoxin by cell cultures of D. lanata are now extensive (3,22,49,63) and are certainly sufficient to show that the process is technically feasible. Again the question of commercial usability is primarily an economic one.

Thus, two systems which have potential for commercial use, can be identified in the published literature. It would be extremely surprising if additional systems were not at reasonably advanced stages of technical development. The state of development of the technology of plant cell cultures is now sufficiently advanced that a commercial process for the production of a biological could be operational within 5 years.

SUMMARY

The biologicals currently produced from plants are generally low molecular weight chemicals such as drug compounds rather than high molecular weight compounds such as proteins. The reasons for considering a fermentation technolgoy based on plant cell cultures have been described. The state of technology of plant cell cultures is sufficiently advanced to suggest that the major questions remaining are economic rather than primarily technical. Limited analysis of the economics of commercial production of biologicals by plant cell cultures shows that a commercial process is certainly feasible.

REFERENCES

1. Akasu, M., H. Itokawa, and M. Fujita. 1976. Biscoclaurine alkaloids in callus tissues of Stephania cepharantha. Phytochemistry (Oxf.) 15:471-473.
2. Alfermann, A.W., D. Merz and E. Reinhard. 1975. Induction of anthocyanin biosynthesis in tissue cultures of Daucus carota. Planta Medica. Suppl. 70-78.
3. Alfermann, A.W., H.M. Boy, P.C. Döller, W. Hagedorn, M. Heins, J. Wahl, and E. Reinhard. 1977. Biotransformation of cardiac glycosides by plant cell cultures. Pgs. 125-141 in W. Barz, E. Reinhard, and M.H. Zenk, eds. Plant tissue culture and its biotechnological application. Springer-Verlag, Berlin. Springer-Verlag, Berlin.
4. Bauch, H.J. and E. Leistner. 1978. Aromatic metabolites in cell suspension cultures of Galium mollugo L. Planta Medica 33:105-123.
5. Bergmann, K. 1960. Growth and division of single cells of higher plants in vitro. J. Gen. Physiol. 43(4):841-851.
6. Butcher, D.N. 1977. Secondary products in tissue cultures. Pgs. 668-693 in J. Reinert and Y.P.S. Bajaj, eds. Plant, cell, tissue, and organ culture. Springer-Verlag, Berlin, New York.

7. Butcher,D.N., and J.D. Connolly. 1971. An investigation of factors which influence the production of abnormal terpenoids by callus cultures of Andrographis paniculata Nees. J. Exp. Bot. 22:314-322.

8. Constabel, F., O.L. Gamborg, W.G.W. Kurz and W. Steck. 1974. Production of secondary metabolites in plant cell cultures. Planta Med. 25:158-165.

9. Döller, G. 1978. Influence of the medium on the production of serpentine by suspension cultures of Catharanthus roseus (l.) G. Don. Pgs. 109-129 in A.W. Alfermann and E. Reinhard, eds. Production of natural compounds by cell culture methods. Proceedings of an international symposium on plant cell culture. BPT-Bericht.

10. Döller, Von G., A.W. Alfermann, and E. Reinhard. 1976. Production of indole alkaloids in tissue cultures of Catharanthus roseus. Planta Medica 30:14-20.

11. Dougall, D.K., G. Whitten and K.W. Weyrauch. Unpublished.

12. Dougall, D.K. 1979. Factors affecting the yields of secondary products in plant tissue cultures. Proceedings of the Fourth Annual College of Biological Sciences Colloquium: Plant Cell and Tissue Culture - Principles and Applications. Ohio State University Press. (in press).

13. Dougall, D.K. 1979. Preservation of germ plasm. University of Tennessee Symposium. "Propagation of Higher Plants Through Tissue Culture: A Bridge Between Applications and Research." (proceedings in press).

14. Dougall, D.K. 1979. Storage of Plant Cell Lines. In E. John Staba, ed. Plant Tissue Culture as a Source of Biochemicals. CRC Press, Inc., Cleveland (in press).

15. Efstratiadis, A., J.N. Vournakis, H. Donis-Keller, G. Chaconas, D.K. Dougall, and F. Kafatos. 1977. End labeling of enzymatically decapped mRNA. Nucleic acids research. 4(12):4165-4174.

16. Farnsworth, N.R. and A.S. Bingel. 1976. Problems and prospects of discovering new drugs from higher plants by pharmacological screening. Pgs. 1-22 in H. Wagner and P. Wolff, eds. New natural products and plant drugs with pharmacological, biological or therapeutical activity. Springer-Verlag. Berlin, Heidelberg, NY.

17. Fowler, M.W. 1977. Growth of cell cultures under chemostat conditions. Pgs. 253-265 in W. Barz, E. Reinhard, and M.H. Zenk, eds. Plant Tissue Culture and Its Bio-technological Application. Springer-Verlag, Berlin.

18. Furuya, T., H. Kojima and K. Syōno. 1971. Regulation of nicotine biosynthesis by auxins in tobacco cellus tissues. Phytochem. 10:1529-1532.

19. Gamborg, O.L., T. Murashige, T.A. Thorpe, and I.K. Vasil. 1976. Plant tissue culture media. In Vitro 12:473-477.

20. Gibbs, J.L. and D.K. Dougall. 1965. The growth of single cells from Nicotiana tabacum callus tissue in nutrient medium containing agar. Exp. Cell Res. 40:85-95.

21. Goldstein, W.E., L.L. Lasure and M.B. Ingle. 1979. Product cost analysis. in E. John Staba, ed. Plant Tissue Culture as a Source of Biochemicals. CRC Press, Inc., Cleveland, OH (in press).
22. Heins, M. 1978. Screening of Digitalis lanata plants and cell cultures for hydroxlation capacity. Pgs. 39-47 in A.W. Alfermann and E. Reinhard eds. Production of natural compounds by cell culture methods. Proceedings of an international symposium on plant cell culture. BPT-Bericht.
23. Hiraoka, N., and M. Tabata. 1974. Alkaloid production by plants regenerated from cultured cells of Datura innoxia. Phytochemistry (Oxf.) 13:1671-1675.
24. Hirotani, M., and T. Furuya. 1977. Restoration of cardenolide synthesis in redifferentiated shoots from callus cultures of Digitalis purpurea. Phytochemistry (Oxf.) 16:610-611.
25. Ikeda, T., T. Matsumoto, and M. Noguchi. 1975. Formation of ubiquinone by tobacco plant cells in suspension culture. Phytochemistry (Oxf.) 15:568-569.
26. Ikeda, T., T. Matsumoto, and M. Noguchi. 1976. Effects of nutritional factors on the formation of ubiquinone by tobacco plant cells in suspension culture. Agri. Biol. Chem. 40:1765-1770.
27. Ikuta, A., K. Syōno and T. Furuya. 1974. Alkaloids of callus tissue and redifferentiated plantlets in Papaveraceae. Phytochemistry (Oxf.) 13:2175-2179.
28. Ikuta, A., K. Syōno, and T. Furuya. 1975. Alkaloids in plants regenerated from Coptis callus cultures. Phytochemistry (Oxf.) 14:1209-1210.
29. Jain, S.C., H. Rosenberg and S.J. Stohs. 1977. Steroidal constituents of Trigonella occulta tissue cultures. Planta Med. 31:109-111.
30. Jhang, J.J., E.J. Staba, and J.Y. Kim. 1974. American and Korean ginseng tissue cultures: growth, chemical analysis, and plantlet production. In Vitro 9:253-259.
31. Kamimura, S., M. Akutsu, and M. Nishikawa. 1976. Formation of thebaine in the suspension culture of Papaver bracteatum. Agric. Biol. Chem. 40:913-919.
32. Kao, K.N. and M.R. Michayluk. 1975. Nutritional requirements for growth of Vicia hajastana cells and protoplasts at very low population density in liquid media. Planta (Berl.) 126:105-110.
33. Kato, A., Y. Shimizu, and S. Nagai. 1975. Effect of initial K_{La} on the growth of tobacco cells in batch culture. J. Ferment. Techol. 58:744-751.
34. Kato, K., Y. Shiozawa, A. Yamada, K. Nishida, and M. Noguchi. 1972. A jar fermentor culture of Nicotiana tabacum L. cell suspensions. Agric. Biol. Chem. 36:899-902.
35. Kato, K., T. Matsumoto, A. Koiwai, S. Mizusaki, K. Nishida, M. Noguchi, and E. Tamaki. 1972. Liquid suspension culture of tobacco cells. Proc. IV IFS: Ferment. Technol. Today 1972:689-695.

36. Kaul. B., S.J. Stohs, and E.J. Staba. 1969. Dioscorea tissue cultures. III. Influence of various factors on diosgenin production by Dioscorea deltoidea callus and suspension cultures. Lloydia 32:347-359.
37. Khanna, P., A. Uddin, and M. Sogani. 1976. Pseudoephedrine from in vivo and in vitro tissue cultures of Ephedra foliata Boiss. Indian J. Pharm. 38:140-141.
38. King, P.J., and H.E. Street. 1973. Growth patterns in cell cultures. Pgs. 269-337 in H.E. Street, ed. Plant Tissue and Cell Culture. (Botanical Monographs, Vol. 11.) University of California Press, Berkeley and Los Angeles.
39. Krikorian, A.D., and F.C. Steward. 1969. Biochemical differentiation: The biosynthetic potentialities of growing and quiescent tissue. Pgs. 227-326 in F.C. Steward, ed. Plant Physiology, A Treatise. Vol. VB. Academic Press, New York.
40. Mandels, M. 1972. The culture of plant cells. Pgs. 201-215 in T.K. Ghose, A. Fiechter, and N. Blakebrough, eds. Advances in Biochemical Engineering 2. Springer-Verlag, Berlin.
41. Misawa, M., H. Tanaka, O. Chiyo, and N. Mukai. 1975. Production of a plasmin inhibitory substance by Scopolia japonica suspension cultures. Biotechol. Bioeng. 17:305-314.
42. Mizukami, H., M. Konoshima, and M. Tabata. 1978. Variation in pigment production in Lithospermum erythrorhizon callus cultures. Phytochemistry (Oxf.) 17:95-97.
43. Muir, W.H., A.C. Hildebrant and A.J. Riker. 1958. The preparation, isolation and growth in culture of single cells from higher plants. Am. J. Bot. 45:589-597.
44. Noguchi, M., T. Matsumoto, Y. Hirata, K. Yamamoto, A. Katsuyama, A. Kato, S. Azechi, and K. Kato. 1977. Improvement of growth rates of plant cell cultures. Pgs. 85-94 in W. Barz, E. Reinhard, and M.H. Zenk, eds. Plant tissue culture and its biotechnological application. Springer-Verlag, Berlin, NY.
45. Ogino, T., N. Hiraoka and M. Tabata. 1978. Selection of high nicotine-producing cell lines of tobacco callus by single-cell cloning. Phytochem. 17:1907-1910.
46. Ohta, S., O. Matsui and M. Yatazawa. 1978. Culture conditions for nicotine production in tobacco tissue culture. Agri. Biol. Chem. 42:1245-1251.
47. Puhan, Z. and S.M. Martin. 1971. The industrial potential of plant cell culture. Prog. Ind. Microbio. 9:13-39.
48. Razzaque, A. and B.E. Ellis. 1977. Rosmarinic acid production in Coleus cell cultures. Planta 137:287-291.
49. Reinhard, E., M. Heins, U. Helmbold, and J. Wahl. 1978. Production of β-Methyldigoxin by cell cultures of Digitalis lanata. Pg. 79 in Abstracts. Fourth international congress of plant tissue and cell culture. Calgary, Alberta Canada.
50. Roller, U. 1978. Selection of plants and plant tissue cultures of Catharanthus roseus with high content of serpentine and ajmalicine. Pgs. 95-108 in A.W. Alfermann and E. Reinhard, eds. Production of natural compounds by cell culture methods.

51. Routien, J.B., and Nickell, L.G. 1956. Cultivation of plant tissue. U.S. Patent. 2, 747, 334.
52. Schulte, U. and M.H. Zenk. 1977. A replica plating method for plant cells. Physiol. Plant. 39:139-142.
53. Staba, E.J. 1977. Tissue culture and pharmacy. Pgs. 694-718 in J. Reinert and Y.P.S. Bajaj, eds. Plant, cell, tissue and organ culture. Springer-Verlag, Berlin, New York.
54. Steck, W., B.K. Bailey, J.P. Shyluk, and O.L. Gamborg. 1971 Coumarins and alkaloids from cell cultures of *Ruta graveolens*. Phytochemistry (Oxf). 10:191-194.
55. Tabata, M. and N. Hiraoka. 1976. Variation of alkaloid production in *Nicotiana rustica* callus cultures. Physiol. Plant. 38:19-23.
56. Tabata, M., H. Mizukami, N. Hiraoka and M. Konoshima. 1974. Pigment formation in callus cultures of *Lithospermum erythrorhizon*. Phytochem. 13:927-932.
57. Tabata, M., H. Yamamoto, N. Hiraoka and M. Konoshima. 1972. Organization and alkaloid production in tissue cultures of *Scopolia parviflora*. Phytochemistry (Oxf.) 11:949-955.
58. Tabata, M., N. Hiraoka, M. Ikenoue, Y. Sano, and M. Konoshima. 1975. The production of anthraquinones in callus cultures of *Cassia tora*. Lloydia 38:131-134.
59. Tabata, M., H. Yamamoto, N. Hiraoka, Y. Marumoto, and M. Konoshima. 1971. Regulation of nicotine production in tobacco tissue culture by plant growth regulators. Phytochemistry (Oxf.) 10:723-729.
60. Tanaka, H., Y. Machida, H. Tanaka, N. Mukai, and M. Misawa. 1974. Accumulation of glutamine by suspension cultures of *Symphytum officinale*. Agric. Biol. Chem. 38:987-992.
61. Tulecke, W., R. Taggart, and L. Colavito. 1965. Continuous cultures of higher plant cells in liquid media. Contrib. Boyce Thompson Inst. 23:33-46.
62. Tyler, V.E., L.R. Brady and J.E. Robbers. 1976. Pharmacognosy. 7th edition Lea and Febiger. Philadelphia.
63. Wahl, J. 1978. Biotransformation of β-Methyldigitoxin by *Digitalis lanata* fermentor cultures. Pgs. 48-59 in A.W. Alfermann and E. Reinhard eds. Production of natural compounds by cell culture methods. Proceedings of an international symposium on plant cell culture. PFT-Bericht.
64. Waller, G.R. and E.K. Nowacki. 1978. Alkaloid biology and metabolism in plants. Plenum Press, New York and London.
65. Westcott, R.J., and G.G. Henshaw. 1976. Phenolic synthesis and phenylalanine ammonia-lyase activity in suspension cultures of *Acer pseudoplatanus* L. Planta. (Berl.) 131:67-73.
66. Wilson, G. and P. Marron. 1978. Growth and anthraquinone biosynthesis by *Galium mollugo* L. cells in batch and chemostat culture. J. Exp. Bot. 29:837-851.
67. Withers, L.A. 1978. Freeze-preservation of cultured cells and tissues. Pgs. 297-306 in T.A. Thorpe, ed. Frontiers of plant tissue culture 1978. International Association for Plant Tissue

Culture. The Bookstore, University of Calgary, Calgary, Alberta Canada.

68. Withers, L.A. and H.E. Street. 1977. The freeze-preservation of plant cell cultures of Digitalis lanata. Pgs. 227-244 in W. Barz, E. Reinhard and M.H. Zenk, eds. Plant tissue culture and its bio-technological application. Springer-Verlag Berlin Heidelberg, New York.

69. Zenk, M.H. 1978. The impact of plant cell culture on industry. Pgs. 1-13 in T.A. Thorpe ed. Frontiers of plant tissue culture 1978. pubs. International Association for Plant Tissue Culture. The Bookstore, University of Calgary, Calgary, Alberta, Canada.

70. Zenk, M.H., El-Shagi and U. Schulte. 1975. Anthaquinone production by cell cultures of Morinda citrifolia. Planta Medica Suppl. 79-101.

71. Zenk, M.H., H. El-Shagi, and B. Ulbrich. 1977. Production of rosmarinic acid by cell-suspension cultures of Coleus blumei. Naturwissenschaften 64:585.

72. Zenk, M.H., H. El-Shagi, H. Arens, J. Stöckigt, E.W. Weiler, and B. Deus. 1977. Formation of the indole alkaloids serpentine and ajmalicine in cell suspension cultures of Catharanthus roseus. Pgs. 27-43 in W. Barz, E. Reinhard, and M.H. Zenk, eds. Plant tissue culture and its bio-technological applications. Springer-Verlag, Berlin, Heidelberg, New York.

73. Kato, A.P., S. Kawazoe, M. Iijima and Y. Shimizu. 1976. Continuous culture of tobacco cells. J. Ferment. Technol. 54: 82-87.

74. Wagner, F. and H. Vogelmann. 1977. Cultivation of plant tissue cultures in bioreactors and formation of secondary metabolites. Pgs. 245-252 in W. Barz, E. Reinhard, and M.H. Zenk, eds. Plant Tissue culture and its bio-technological applications. Springer-Verlag, Berlin.

75. Veliky, I.A. 1977. Effect of pH on tryptophol formation by cultured Ipomoea sp. plant cells. Lloydia 40:482-486.

USE OF RECOMBINANT DNA TECHNOLOGY FOR THE PRODUCTION OF POLYPEPTIDES

Walter L. Miller, M.D.

Endocrine Research Division and the Department of Pediatrics, University of California, San Francisco

I. INTRODUCTION

There are many polypeptides of biologic and medical interest which have not been fully studied because they cannot be obtained in adequate quantity with sufficient purity and economy. Recent advances in recombinant DNA technology now give promise of *in vivo* synthesis in bacteria of a wide variety of polypeptide hormones, specific immunoglobins, enzymes and other proteins. This presentation will first review the procedures for constructing chimeric microorganisms containing the DNA coding for a eucaryotic protein, then discuss the problems and experience in obtaining such proteins from DNA cloned in procaryotic cells.

Both eucaryotic and procaryotic cells synthesize proteins in comparable manners: DNA is transcribed by a complex polymerase into RNA, which may require further processing before it can be used as a messenger RNA (mRNA); mRNA is translated into protein on ribosomes, using amino-acylated transfer RNAs as the interface between nucleic acid and polypeptide; the resulting polypeptide may then undergo further processing before being utilized by the cell. As these steps are common to all organisms, one may hypothesize that it should be possible to construct chimeric microorganisms containing DNA for a desired eucaryotic polypeptide, and that proper arrangement of this DNA in the host cell should result in the production of the coded protein.

In order to modify microorganisms so that they may produce a desired polypeptide, five steps are necessary. First, DNA sequences coding for the protein are prepared. Secondly, this gene must be covalently linked to the DNA of a suitable vehicle, such as the small, circular, non-chromosomal DNA's termed plasmids, or the DNA of a bacteriophage. The gene for the protein then travels as a passenger with the vehicular DNA when the vehicle infects a suitable recipient host cell. This infection of the host cell is the third step, transformation. Fourth, the transformed host cell must replicate the infecting DNA faithfully so that all daughter cells will carry the gene for the protein. Finally, the transformed cells must transcribe the foreign DNA into mRNA, and the cellular machinery for protein synthesis must accept and translate this mRNA.

While the outlines of molecular biology are the same for both nucleated eucaryotic and lower procaryotic cells, a number of differences exist between these classes of organisms which could interfere with bacterial production of eucaryotic proteins. Various organisms exhibit codon preference, i.e., not all the triplet codons for a particular amino acid are utilized in equal numbers. For example, the amino acid leucine may be specified by six different codons, but CUG is used for 19 out of 31 residues in rat pre-growth hormone (1) and in 11 out of 21 residues in human chorionic somatomammotropin (2); however, the same codon is used in only 16 of 143 residues in bacteriophage MS2 (3). Similarly, promoter sequences and ribosome binding sites also differ between eucaryotic and procaryotic organisms (4-6). Each of these types of differences, and possibly others, has the potential for altering the efficiency of transcription and translation of eucaryotic gene sequences which have been inserted into a procaryotic genome by recombinant DNA technology.

The strategy which has been most widely employed in recombinant DNA technology is outlined in Figure 1. DNA, foreign to the host microorganism and which codes for the polypeptide of interest, is isolated and cut with one of several restriction endonucleases. A restriction endonuclease is a highly specific enzyme which cuts double stranded DNA at a uniquely specified sequence of four to six nucleotides. These enzymes recognize mirror image sequences of DNA (pallindromes) and cut them along an axis of symmetry, commonly generating short segments of single stranded DNA which are complementary to each other. Such complementary ends of DNA strands are termed cohesive termini, or "sticky ends" because they have the ability to recognize and adhere to complementary sticky ends by base pairing. A cloning vehicle, in this example a plasmid, is chosen with a single restriction site for the enzyme used to cut the foreign DNA. The plasmid is cut with the enzyme and mixed with the foreign DNA. The complementary sticky ends recognize each other and align by hydrogen bonding, inserting the foreign DNA into the plasmid. The ends are then cemented by DNA ligase. The resulting new chimeric plasmid, carrying the gene for the desired protein, is then put into bacteria (usually a derivative of E. coli strain K12). Plasmids

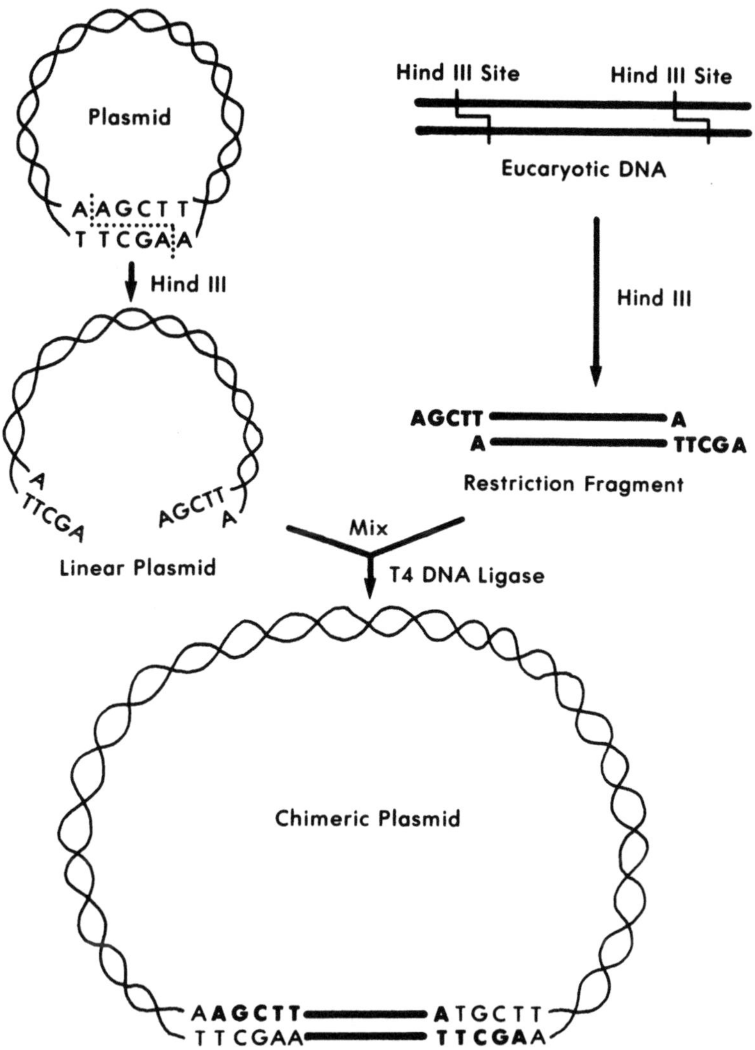

Figure 1. Strategy for cloning foreign DNA in a microorganism using a plasmid vector. Foreign DNA is cut with a restriction enzyme such as Hind III, yielding fragments with self-complementary cohesive termini. A plasmid cloning vector having a single restriction site for Hind III is also cut with the enzyme. The cut ("linear") plasmid is mixed with the foreign DNA. The complementary cohesive Hind III termini of plasmid and foreign DNA align by hydrogen-bond base pairing, and the DNA strands are sealed with DNA ligase. This "chimeric plasmid" may be replicated in transformed bacteria.

which have been taken up by the bacteria now behave as benign obligate intracellular parasites, replicating independently of the host chromosomal DNA, and have full access to its protein synthetic machinery. Successfully transformed bacteria can be selected on agar plates if the plasmid cloning vehicle also confers antibiotic resistance to the bacteria.

II. MEANS OF OBTAINING DNA SEQUENCES

Progress made in many laboratories in the past five years has made available a wide variety of restriction enzymes, ligation procedures, cloning vehicles, and hosts. This technology has been limited in techniques for preparing DNA for cloning, in selecting the proper colonies of chimeric microorganisms, and in obtaining expression of the cloned DNA in the microorganisms. The principal difficulty in obtaining human DNA has been ensuring its purity. Because of speculations concerning the safety of recombinant DNA research, regulations are delineated in a series of NIH guidelines (7), wherein the source and purity of the DNA, the cloning vehicles, and the recipient microorganisms are specified so as to provide biologic containment. Various levels of physical containment in specially built and certified laboratories are also specified by these guidelines. However, it must be emphasized that there is no evidence that these containment measures are necessary. DNA sequences for cloning can be generated in three ways: they can be created synthetically by elaborate but conventional organic chemistry; they can be isolated from the chromosomal DNA of an organism; or they can be enzymatically copied from purified mRNA. Each of these techniques has different strengths and weaknesses which would then dictate the best approach to cloning the DNA for a specific protein.

a. Chemical Synthesis

Chemically synthesized DNA offers formidable advantages for many recombinant DNA experiments. Chemical synthesis can yield a product of great purity, without the possibility that the DNA sequences for other genes have been co-isolated with the DNA sequences to be cloned. Secondly, as the DNA is being chemically synthesized, its precise nucleotide sequence must be designed by the experimenter. This permits selection of triplet codons favored by the recipient host microorganism and permits the creating of designed proteins which do not exist in nature. Such modified proteins have already been created by protein chemistry, resulting in pharmacologically useful derivatives of vasopressin, oxytocin, and gonadotropin releasing hormone. However, chemical synthesis is complicated and may not be able to produce long segments of DNA efficiently. Furthermore, as short polypeptides are rapidly degraded by the host bacteria, successful production of these peptides has been achieved by covalent linkage to large proteins which must be removed without harming the peptide coded by the cloned DNA. Despite these limitations, this approach has been used to produce somatostatin, a 14 amino acid peptide, in bacteria (8) (Figure 2). The plasmid pBR322 was modified by insertion of the E. coli lac operon along with the structural gene for β-galactosidase. A synthetic gene coding for somatostatin and containing an extra triplet codon for a amino-terminal methionine residue is joined to the β-galactosidase gene at an Eco Rl restriction enzyme site. The synthetic somatostatin gene also contains two different nonsense or "stop" codons following the codon for the carboxy-terminal cystine residue, and is joined to pBR322 at a Bam I site. Since the lac repressor gene is absent in this plasmid, and since the lac repressor of the host E. coli RRl chromosomal DNA cannot produce enough repressor molecules to bind to the operator genes of all plasmid copies present in the cell, the cell essentially becomes constitutive for the production of β-galactosidase. Each molecule of β-galactosidase coded by the plasmid will then have the 14 amino acids of somatostatin linked to its carboxy terminus by an extra methionine residue. This hybrid protein molecule is then cleaved at its methionine residues with cyanogen bromide to yield active somatostatin and fragments of β-galactosidase.

b. Isolation of Native DNA Fragments

The most direct approach to obtaining long sequences of DNA for cloning is the isolation of native gene fragments. This is the approach used in early experiments cloning bacterial DNA (9) and DNA's coding for ribosomal (10) and transfer RNAs (11). This direct approach offers the advantage of permitting the cloning of a whole natural gene with its attendant regulatory regions; furthermore, only the capacity of the cloning vehicle limits the length of DNA which can be cloned by this procedure. Isolation of a whole natural gene makes this a very powerful technique for studying gene structure and regulation, but makes it a poor technique for constructing chimeric microorganisms which can produce eucaryotic proteins. Procaryotic host cells might recognize eucaryotic control sequences and enzyme binding sites poorly. Furthermore, eucaryotic genes are often interrupted by "intervening" sequences (12,13). Intervening sequences are segments of DNA, often of considerable length, which interrupt the coding regions of certain genes. These sequences appear to be transcribed into heterogeneous nuclear RNA, which is then processed into mRNA by excision of these sequences (14). As procaryotic organisms do not process their mRNA in this matter, they may not be able to process the RNA transcribed from a cloned eucaryotic gene containing intervening sequences, and hence would not express this genetic material as meaningful protein. Another limitation of this procedure is the need for a relatively pure probe, a radioactive complementary DNA or mRNA which can be used to assay for the presence of the desired sequences.

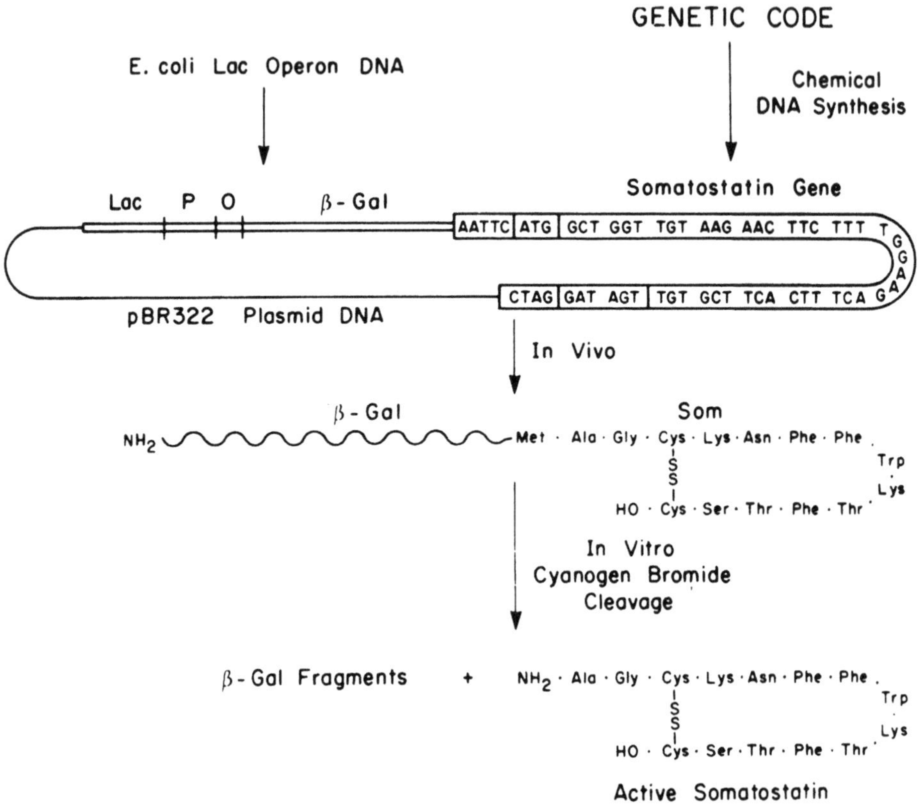

Figure 2. Modification of plasmid pBR322 for the production of somatostatin. A fragment of E. coli DNA containing the lac operon, including the promoter and operator sites, and a portion of the structural gene for β-galactosidase had been inserted into the plasmid pBR322. A chemically synthesized double-stranded segment of DNA is attached to the β-galactosidase gene at an Eco RI site ((G)AATTC) and to the pBR322 at a Bam I site ((C)CTAG(G)). The sequences coding for somatostatin are linked to the Eco RI site by a methionine codon (ATG), and to the Bam I site by two nonsense (stop) condons (GAT AGT). When inserted into E. coli, the β-galactosidase and somatostatin sequences are transcribed under control of the lac operon, yielding a large, hybrid protein molecule. Digesttion of this protein with cyanogen bromide cleaves all the methionine residues, fragmenting the β-galactosidase and liberating active somatostatin. (Reproduced by permission from Itakura et al., Science 189:1056, 1977) (8). Copyright 1977 by The American Association for the Advancement of Science.

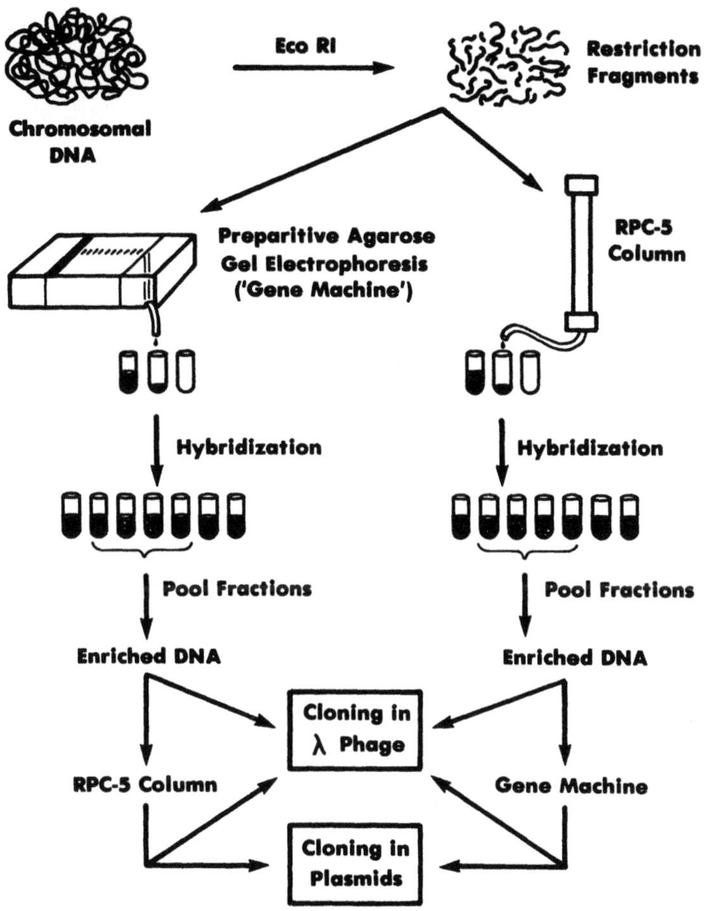

Figure 3. Cloning of native gene fragments. DNA is prepared from nuclei and cut with a restriction fragment such as Eco RI. The fragments are roughly separated by size and charge either by agarose gel electrophoresis or reverse-phase chromatography. Aliquots of collected fractions are assayed for the desired sequences by nucleic acid hybridization and the appropriate fractions are pooled. This partially purified DNA may be cloned directly or purified further.

The general strategy for cloning native DNA sequences is shown in Figure 3. Chromosomal DNA is cut with a restriction endonuclease such as Eco Rl chosen to give relatively large restriction fragments. Digestion of human DNA with this enzyme yields approximately 10^6 unique fragments. These may be partially separated according to size and charge by preparative agarose gel electrophoresis (15) or reverse-phase column chromotography (16). Fractions of the eluate are collected and aliquots are assayed for the presence of the subject DNA sequences by nucleic acid hybridization (17). This assay requires the use of a radioactively labeled probe, usually a strand of DNA synthesized on a template of mRNA by reserve transcriptase(18), and hence complementary to it. Fractions containing the relevant DNA sequences are then pooled and the DNA precipitated. This DNA is enriched about 100-fold for the desired sequences, but still contains a vast excess of other DNA. The enriched DNA can be purified further by either RPC-5 chromatography or preparative agarose gel electrophoresis (whichever was not used first) prior to cloning in plasmids as described earlier. Alternatively, it may be inserted into bacteriophage lambda and cloned when the recombinant phage infects suitable E. coli, taking advantage of the greatly increased number of recombinant molecules which may be screened with this technique (19). Recently, the native gene for ovalbumin (20),and a seven kilobase fragment of mouse DNA containing the gene for β globin, at least one insert, and extensive regions on both sides of the structural gene (21) have been cloned using this approach. Figure 4 shows the map of the ovalbumin gene as determined by Dugaiczyk et al., derived from detailed restriction enzyme analysis of several Eco Rl fragments purified and cloned as described above. The startling lesson from this map is that a structural gene over 7,000 bases long is used to code for an mRNA of only 1,859 bases, thus the sequences actually coding for protein may constitute a small fraction of a structural gene.

c. Reverse Transcription

The third strategy for obtaining DNA sequences for cloning is to obtain the RNA for the protein of interest, make a strand of DNA complementary to it by reverse transcription, then clone this complementary DNA (cDNA). The advantages of this approach are obvious: first, there are no intervening sequences, as cDNA is copied from mRNA, which is free of intervening sequences. Second, a nucleic acid probe is not crucial to this approach. Third, the cDNA can be highly labelled with radioactive nucleotides thereby greatly facilitating its handling and anlysis. These advantages may be offset somewhat by inefficient mRNA isolation and incomplete reverse transcription, so that the cloned cDNA may be only a fragment of the gene. Secondly, although it is possible to obtain mRNA and cDNA enriched for a particular sequence, it is almost impossible to obtain it in an absolutely pure form. As a result, cloning cDNA is usually a "shotgun experiment" where bacterial colonies containing the desired chimeric plasmids must be selected from colonies containing extraneous material. This raises a third problem, detecting the bacterial clones containing the sequences of interest. Despite these difficulties, cDNA cloning has produced chimeras harboring genes for growth hormone, insulin, ACTH, chorionic somatomammotropin, ovalbumin, dihydrofolate reductase, immunoglobulin chains, and others. Figure 5 illustrates this technique: Tissue which synthesizes a large quantity of the desired protein, and hence is rich in its mRNA, is chosen and polyadenylated mRNA is prepared from it by standard techniques. Proper choice of tissue or physiologic prepa-

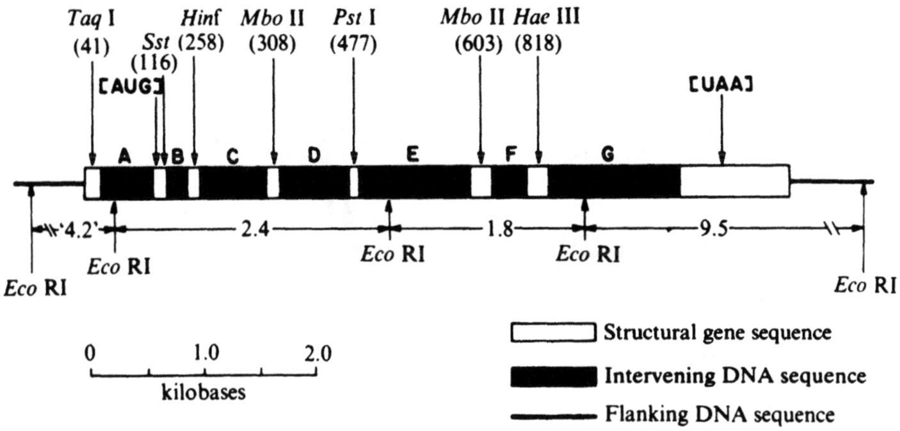

Figure 4. Structure of the ovalbumin gene. The ovalbumin-coding sequences totaling 1859 bases, are divided into seven pieces by long segments of DNA (intervening sequences or "introns"), so that the whole gene is over 7,000 bases long. Intervening sequences appear to be transcribed into RNA, partially explaining the role of HnRNA precursors to mRNA, but are not translated into protein. (Reproduced by permission from Dugaiczyk et. al., Nature 274:328, 1978) (20).

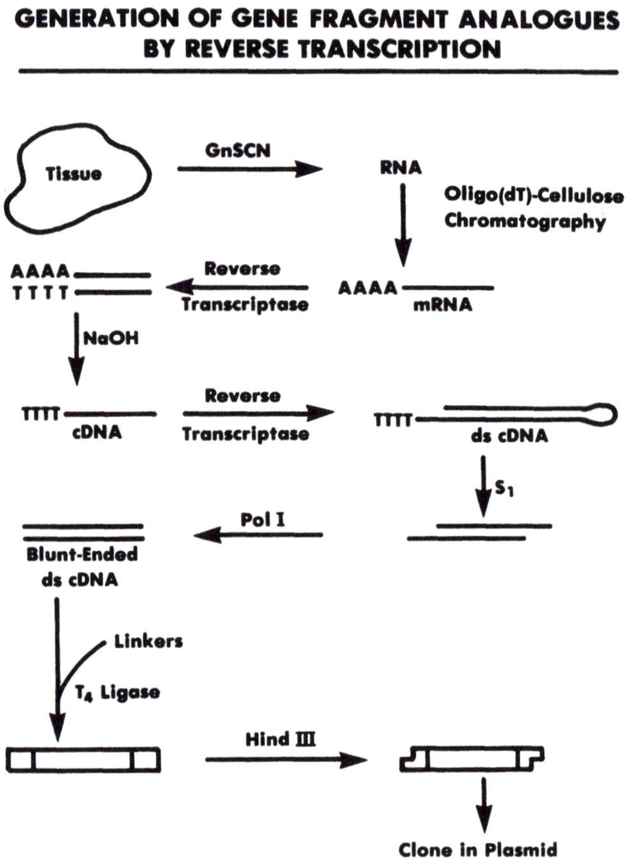

Figure 5. Preparation of complementary DNA for cloning. RNA is prepared by guanidine thiocyanate extraction from tissue or cells containing a large amount of mRNA coding for a protein of interest. Polyadenylated mRNA is selected by affinity chromatography on oligo (dT)-cellulose and copied into DNA by avian myeloblastosis virus reverse transcriptase. After alkali digestion of the mRNA template, the single-stranded cDNA is reverse transcribed into a hairpin-shaped double stranded cDNA. The hairpin is opened and the single stranded ends are partially digested with S_1 nuclease, and the remaining single stranded ends are filled in by DNA polymerase I to yield blunt-ended cDNA. Synthetic decanucleotide "linkers" containing the recognition sequence of the restriction enzyme Hind III are attached to the blunt-ended cDNA by DNA ligase from bacteriophage T_4. The cDNA is then digested with Hind III to yield cohesive termini suitable for ligation into a plasmid as shown in Figure 1.

ration can greatly increase the yield of a specific mRNA: for example, ovalbumin mRNA constitutes 50% of the mRNA in the oviduct of an estrogen-treated chicken, but only 0.01% of the mRNA in the oviduct before estrogen treatment (22). Similarly, treatment of the cultured rat pituitary tumor cell line GC with thyroxine and dexamethasone resulted in an increased quantity of growth hormone mRNA to about 10% of the total (23). A greater purification can be obtained by preparing polyribosomes and selecting those carrying the desired mRNA by immune precipitation of the attached nascent protein, but this yields very little mRNA. Polyadenylated mRNA is then copied into cDNA by reverse transcriptase, an enzyme derived from certain RNA tumor viruses (24,25). After digesting the mRNA template with alkali, a second strand of cDNA is made with the same enzyme. This double stranded cDNA must be processed further. It can be digested with a restriction enzyme which will cut its length, but also give sticky ends which may be ligated into a plasmid cut with the same enzyme. Alternatively, the hairpin loop formed by the reverse transcriptase may be opened enzymatically with S1 nuclease and the unpaired ends filled in by DNA polymerase I to yield "blunt-ended cDNA." This full length cDNA can then be cloned using "linkers" synthetic double-stranded polynucleotides 8-10 base pairs long, which contain the recognition sequence for a restriction enzyme (26). The linkers are attached to the blunt-ended cDNA with T_4 DNA ligase (27) and digested with the restriction enzyme to yield sticky ends. This cDNA can then be cloned in a plasmid opened with the same enzyme. The cDNA may also be cloned by "tailing" it with homopolymeric dC, added to its 3' ends with terminal transferase. Opened plasmid tailed with dG is mixed with the dC-tailed cDNA under annealing conditions so that stable dC·dG base pairs form, thus recircularizing the plasmid with the cDNA insert (28,29).

III. HOST-VEHICLE SYSTEMS AVAILABLE

Once we have prepared the DNA, a suitable cloning vehicle must be chosen. Three distinct possibilities exist: bacterial plasmids, λphage, and animal viruses, of which only the first two are in general use. Bacterial plasmids, also known as R-factors or episomes, are small circular pieces of extra-chromosomal DNA found in a wide variety of gram-negative bacteria. They have long been known to carry genes for resistance to various antibiotics and to be transferable between bacteria by cell-to-cell contact. Plasmids were the first vehicles used for cloning (10,30), and remain the most widely used technique. A number of plasmids have been specifically "built" for use as vehicles for recombinant DNA research (31,32). These are derived from the naturally occurring plasmid Col E1 and created by restriction endonuclease excision of unwanted DNA sequences and insertion of others. The plasmid vehicle with its passenger DNA may be assimilated into bacteria which have been rendered permeable to extracellular DNA by pretreatment with calcium chloride. Because of present requirements for biological containment, recipient bacteria have been devised by genetic manipulation so that they can survive only in uniquely enriched environments which cannot exist in nature. For example, E. coli χ1776, which requires a medium containing both thymidine (for nucleic acid synthesis) and diaminopimelic acid (a constituent of the cell wall) has been derived from strain K12, which itself cannot survive in the human gut (33). Such host-vector systems have been used in hundreds of successful cloning experiments, but remain limited by the difficulty of screening large numbers of clones. To this end, a number of modifications have been built into the E. coli phage λ (34-38) which permit screening of thousands of clones (19). In one case, expression of cloned globin DNA was obtained in a λphage system when cloning in a plasmid system had failed to produce detectable protein (39). Other cloning vehicle systems such as somatic cells infected with simian virus 40 have been used (40,41) and may soon be practical for the *in vitro* production of many biologically useful polypeptides.

IV. EXPRESSION OF CLONED DNA

The earliest evidence that recombinant DNA could be expressed in transformed E. coli was the expression of genes native to the host E. coli itself, such as the *lac* operator (42,43). Transcribed RNA complementary to cloned DNA was detected by several investigators, (10, 44, 45) but much of this transcription was regarded as "gratuitous" as the protein products of the cloned genes could not be detected. Using the novel system of "minicells" (46), small fragments of E. coli cells containing all normal cellular elements except the chromosomal DNA, Megaher *et al* demonstrated production of some yeast and drosophila proteins from cloned DNA, but could not detect expression of cloned mouse mitochrondrial DNA(47). As discussed in section IIa, expression of the cloned synthetic somatostatin gene has been achieved, but as this entails synthetic DNA, it does not answer the question of whether procaryotic organisms can be made to transcribe and translate eucaryotic genes to yield useful proteins. Production of yeast enzymes has been achieved in both λphage (48) and plasmid (49) systems, but yeasts may not be so dissimilar from procaryots as are vertebrates. Recently however, rat proinsulin (28) mouse dihydrofolate reductase (29), ovalbumin (50), and rat growth hormone (51), have been synthesized in bacteria harboring recombinant plasmids. As an example of this technology, we shall now review the rat growth hormone work in more detail. Using the strategy of reverse transcription (see section IIc above) the entire coding sequence for pre-RGH, the 3' untranslated region of the mRNA, and most of the 5' untranslated region was cloned in the plasmid pBR322 to form the new plasmid pRGH1 (1). The 800 base-pair rat growth hormone gene was then prepared in quantity in pRGH1, excised with Hind III, and used to prepare the "expression plasmid" as shown in

Figure 6. The RGH gene prepared from pRGH1 was recloned in the Hind III site of pMB9. As the sole Hind III site of pMB9 lies a few bases into the gene for tetracycline resistance, cloning a segment of DNA in this site results in the plasmid conferring resistance only to low concentrations of tetracycline, thus providing a selection technique for the successful recombinants (52). A strain of recombinants having the poly dA·dT end of the gene (i.e. the carboxy-terminus of RGH) oriented toward the Eco Rl site was chosen for further treatment. This pMB9-RGH plasmid was then digested with the enzymes Pst I and Bam I. Pst I cuts the RGH gene precisely between the codons for amino acids -24 and -23 of the pre-sequence of RGH (1). The smaller of the two resulting DNA fragments, containing most of the TET gene but little of the RGH gene, was discarded. The plasmid pBR322 was also cut with Pst I and Bam I. Pst I cuts the β-lactamase gene of pBR322 precisely between amino acids 182 and 183 (53), hence both Pst I cleavages occurred at the beginning of codons in the respective genes so that these cleavages were in the same "reading frame." The smaller of the two fragments of pBR322, containing most of the TET gene, the Hind III and Eco Rl sites, and part of the AMP gene was then ligated to the larger of the two fragments of pMB9-RGH. The resulting new plasmid, termed the "expression plasmid" (pEx-RGH), now contains a fully reconstituted and functional TET gene, part of the β-lactamase (AMP) gene (coding for amino acids -23 to 181), and most of the RGH gene (coding for amino acids -24 to 190). Figure 7 depicts the details of the interface between the RGH and AMP genes; the correct reading frame for translation is maintained for both genes. The resulting hybrid RGH-AMP gene should code for a protein of 395 amino acids containing the amino-terminal 181 amino acids of β-lactamase connected in an ordinary peptide linkage to 214 residues of rat pre-growth hormone.

The expression plasmid and pBR322 were then grown in E. coli minicells and the proteins produced from these plasmids analyzed by SDS-acrylamide gel electrophoresis. Both pre-β-lactamase and β-lactamase were synthesized in minicells containing pBR322, but neither was synthesized in minicells containing the expression plasmid pEX-RGH. Minicells containing pEX-RGH produced a new protein with an apparent molecular weight of 46,000, in good agreement with the 44,317 molecular weight of the predicted 395 amino acids in the hybrid protein.

The presence of growth hormone synthesized by E. coli containing pEx-RGH is also inferred from immunologic studies. Bacterial colonies were lysed on agar plates with chloroform and overlain with polyvinyl discs coated with monkey anti-RGH (54). (Figure 8). Rat growth hormone binding to antibody on the discs is detected with highly purified ^{125}I anti-RGH IgG by autoradiography. This technique readily distinguishes bacterial colonies harboring pEx-RGH from colonies harboring pBR322, or colonies harboring no plasmid.

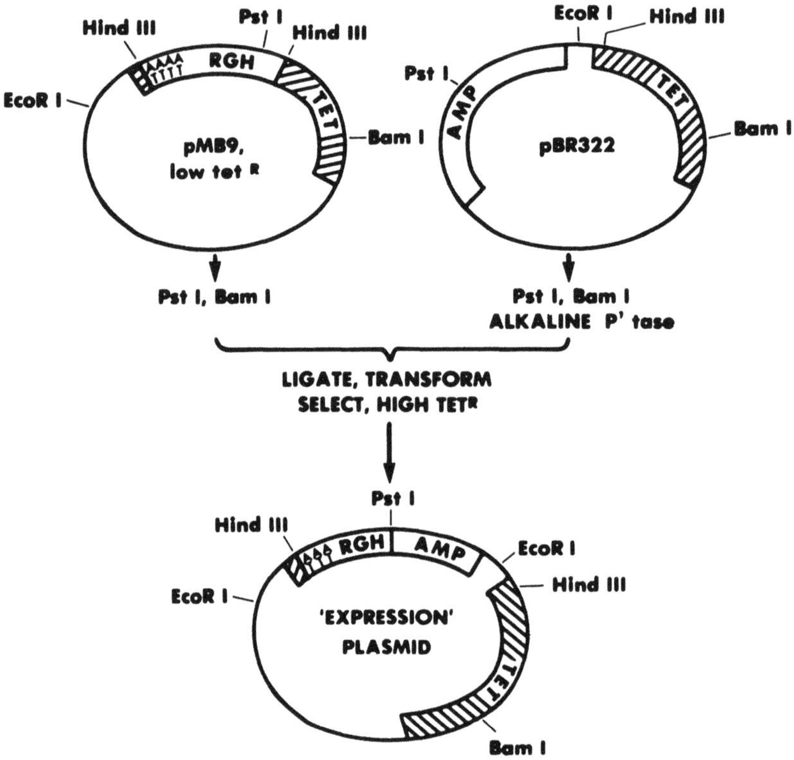

Figure 6. Construction of the expression plasmid pExRGH. An 800-nucleotide gene fragment coding for rat growth hormone, its entire 3' untranslated region, and most of the 5' untranslated region was first cloned in pRGH1, then re-cloned in the Hind III site of pMB9 oriented with its poly dA·dT end toward the Eco RI site. The Hind III site of pMB9 lies a few nucleotides into the TET gene so that pMB9-RGH imparts resistance to only very low concentrations of tetracycline, permitting selection of bacterial colonies containing chimeric plasmids. The pMB9-RGH was cut with Pst I and Bam I and the smaller of the resulting fragments, containing much of the TET gene, but only a few bases of the RGH gene, was discarded. The plasmid pBR322 was similarly cut with Pst I and Bam I. The smaller pBR322 fragment, containing part of the AMP gene, the Eco RI and Hind III sites, and most of the TET gene was then ligated into the larger fragment of pMB9-RGH to yield the expression plasmid pExRGH. The resulting plasmid thus contains the RGH gene linked to the AMP gene at a Pst I site, and a completely reconstituted TET gene.

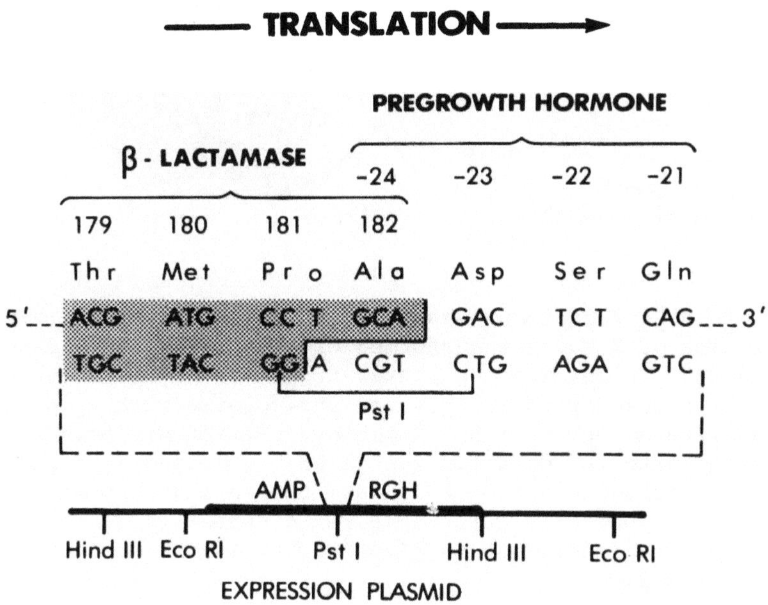

Figure 7. Detail of the Pst I site of pExRGH. Pst I recognizes a 6-nucleotide sequence and generates cohesive termini 4 nucleotides long, cutting pBR322 in the middle of the β-lactamase gene which imparts ampicillin resistance, and the RGH gene in the pregrowth hormone region. Because the complete nucleotide sequences of the β-lactamase and RGH genes were known, it was known that this ligation at the Pst I site would maintain the correct reading frame for both genes. This is easily seen as the enzyme happens to cut between the codons for alanine and aspartic acid.

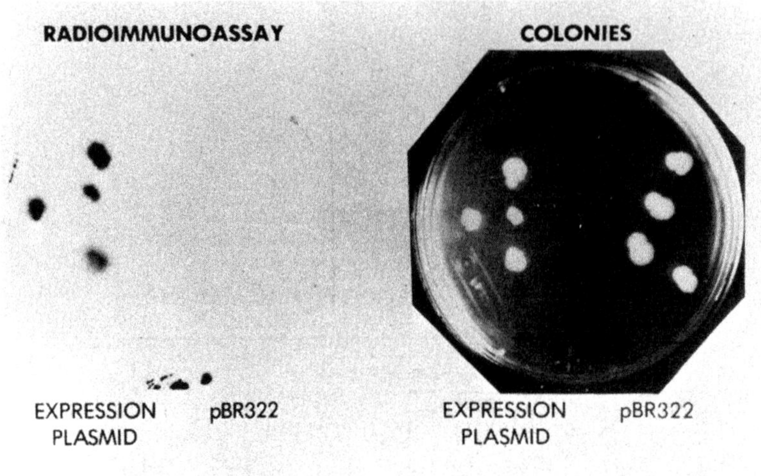

Figure 8. In situ radioimmunoassay of E. coli containing pExRGH. Right: Colonies of E. coli transformed with pExRGH (left side of plate) or pBR322 (right side of plate) are indistinguishable. The cells were then lysed with $CHCl_3$ vapor and covered with a polyvinyl disc coated with monkey anti-RGH antibody. Rat growth hormone binding to the disc was detected with highly purified ^{125}I anti-RGH. Left: Autoradiograph of polyvinyl disc assay of plate shown on right. Immunologically detectable rat growth hormone was present in E. coli transformed with pExRGH but not in E. coli transformed with pBR322.

Labeling of pEx-RGH colonies could be inhibited by the presence of excess growth hormone, but not by the presence of normal monkey serum, ^{125}I normal-monkey serum or calf serum, demonstrating the assay's specificity. Using this procedure, 8.0 ng of exogenous RGH added to a bacterial plate could be detected, but 0.8 ng could not.

V. SPECULATION

Recombinant DNA technology appears to have the potential for revolutionizing biology and pharmacology. Individual phenotypic traits from one species may now be transferred to other species by the transfer of specific DNA coding for those traits. The transcription and translation of these recombinant molecules may be utilized for the production of various polypeptides *in vivo*, promising an inexpensive source of pharmacologically useful proteins which formerly were unavailable. Insulin-requiring diabetics need no longer depend on the diminishing supply of beef and pork insulins. The supply of human growth hormone will no longer depend on extraction of cadaveric human pituitaries, and will thus become available for experimental and therapeutic use in a wide variety of growth and metabolic disorders. The other protein hormones will also become available, ending dependency on animal sources, expensive solid-state protein synthesis, or inadequate alternative therapies for treating deficiencies of these hormones. With the recent cloning of the gene for human growth hormone, it appears the era of recombinant DNA pharmaceuticals is immediately imminent. Virtually any polypeptide could, in theory, be produced on an industrial scale in this fashion: immunoglobulins, vaccines, antitoxins, clotting factors, kinins, etc.

Some of the techniques described above for preparing DNA for cloning may be combined. Naturally occurring genes for hormones or antibodies could be modified by coupling to segments of synthetic DNA to produce new proteins with pre-designed properties and affinities. Chemical synthesis of genes is already showing the way to designed proteins. This may provide a new approach to the study of structure-function relationships in proteins, and offer pharmacologists the opportunity to modify "functional groups" on protein as they do on other pharmaceuticals.

In addition to using recombinant DNA as a means of synthesizing proteins for use, new and more useful organisms may eventually be designed. Grain crops may be made capable of nitrogen fixation and may be designed to have balanced amino acid compositions, thus improving the world's nutrition. Bacteria may be built to become scavengers of oil spills and other pollutants. The potential of this technology may, in fact, be so great that it will usher in a biologic revolution comparable to the industrial revolution of 200 years ago.

SUMMARY

DNA sequences corresponding to specific genes may be prepared by chemical synthesis, isolation of naturally occurring DNA, or reverse transcription. Such DNA may then be inserted into vectors such as plasmids or bacteriophages which carry the DNA into bacterial cells. Although significant differences exist in the basic molecular biology of eucaryotic and procaryotic organisms, these differences do not constitute absolute barriers to the expression of eucaryotic genes in bacteria. Several eucaryotic proteins, including insulin, growth hormone, ovalbumin, dihydrofolate reductase and somatostatin have been produced in bacteria. The use of chimeric microorganisms harboring recombinant DNA offers a completely new approach to the production of biologically useful polypeptides.

ACKNOWLEDGEMENT

The author wishes to thank John D. Baxter, Howard M. Goodman and Joseph A. Martial for critical reading of the manuscript. The author is a recipient of a clinical investigation award from the National Institute of Arthritis, Metabolism and Digestive Diseases.

REFERENCES

1. Seeburg, P.H., Shine, J., Martial, J.A., Baxter, J.D., and Goodman, H.M. Nucleotide sequence and amplification in bacteria of the structural gene for rat growth hormone. Nature 270:486-494. 1977.
2. Shine, J., Seeburg, P.H., Martial, J.A., Baxter, J.D., and Goodman, H.M. Construction and analysis of recombinant DNA for human chorionic somatomammotropin. Nature 270:494-499. 1977.
3. Fiers, W., Contreras, R., Duerinick, F., Haegeman, G., Iserentant, D., Merregaert, J., Joy, W.M., Molemans, F., Raemaekers, A., Van den Berghe, A., Volkaert, G., and Ysebaert, M. Complete nucleotide sequence of bacteriophage MS2 RNA: primary and secondary structure of the replicase gene. Nature 260: 500-507. 1976.
4. Shine, J. and Delgarno, L. Identical 3'-terminal octanucleotide sequence in 18S ribosomal ribonucleic acid from different eukaryots. Biochem. J. 141:609-615. 1974.
5. Shine, J. and Delgarno, L. Terminal-sequence analysis of bacterial ribosomal RNA, correlation between the 3'-terminal-polypyrimidine sequence of 16-S RNA and translational specificity of the ribosome. Eur. J. Biochem. 57:221-230. 1975.
6. Steitz, J.A., and Jakes, K. How ribosomes select initiator regions in mRNA: base pair formation between the 3' terminus of 16 S rRNA and the mRNA during initiation of protein synthesis in Escherichia coli. Proc. Natl. Acad. Sci. 72:4734-4738. 1975.
7. Recombinant DNA research guidelines. Federal Register 41: 27902-27943 (1976); Recombinant DNA Research Revised guidelines. Federal Register 43: 60080-60131(1978).
8. Itakura, K., Hirose, T., Crea, R., Riggs, A.D., Heyneker, H.C. Bolivar, F., and Boyer, H.W. Expression in Escherichia coli of a chemically synthesized gene for human somatostatin. Science 198:1056-1063. 1977.
9. Chang, A.C.Y., and Cohen, S.N. Genome construction between bacterial species *in vitro*: replication and expression of Staphylococcus plasmid genes in Escherichia coli. Proc. Natl. Acad. Sci. 1030-1034. 1974.
10. Morrow, J.F., Cohen, S.N., Chang, A.C.Y., Boyer, H.W., Goodman, H.M., and Helling, R.B. Replication and transcription of eukaryotic DNA in Escherichia coli. Proc. Natl. Acad. Sci. 71: 1743-1747. 1974.

11. Yeu, P.H., Sodja, A., Cohen, Jr., M.C., Conrad, S.E., Wu, M., Davidson, N., and Ilgen, C. Sequence arrangement of tRNA genes on a fragment of Drosophila melanogaster DNA cloned in E. coli. Cell 11:763-777. 1977.
12. Breathnach, R., Mandel, J.L., and Chambon, P. Ovalbumin gene is split in chicken DNA. Nature 270:314-319. 1977.
13. Chow, L.T., Gelinas, R.E., Broker, T.R., and Roberts, R.J. An amazing sequence arrangement of the 5' ends of adenovirus-2 messenger RNA. Cell 12:1-8. 1977.
14. Roop, D.R., Nordstrom, J.L., Tsai, S.Y., Tsai, M.-J., and O'Malley, B.W. Transcription of structural and intervening sequences in the ovalbumin gene and identification of potential ovalbumin mRNA precursors. Cell 15:671-685. 1978.
15. Polsky, F., Edgell, M.W., Seidman, J.G., and Leder, P. High capacity gel preparative electrophoresis for purification of fragments of genomic DNA. Analyt. Biochem. 87:397-410. 1978.
16. Hardies, S.C., and Wells, R.D. Preparative fractionation of DNA restriction fragments by reversed phase column chromatography. Proc. Natl. Acad. Sci. 73:3117-3121. 1976.
17. Southern, E.M. Detection of specific sequences among DNA fragments separated by gel electrophoresis. J. Mol. Biol. 98:503-517. 1975.
18. Verma, I.M., Temple, G.F., Fan, H., and Baltimore, D. In vitro synthesis of DNA complementary to rabbit reticulocyte 10S RNA Nature 235:163-167. 1972.
19. Benton, W.D., And Davis, R. Screening λgt recombinant clones by hybridization to single plaques in situ. Science 196: 180-182. 1977.
20. Dugaiczyk, A., Woo, S.L.C., Lai, E.C., Mace, Jr., M.L., McReynolds, S.L., and O'Malley, B.W. The natural ovalbumin gene contains seven intervening sequences. Nature 274:328-333. 1978.
21. Tilghman, S.M., Tiemeier, D.C., Polsky, F., Edgell, M.H., Seidman, J.G., Leder, A., Enquist, L.W., Norman, B., and Leder, P. Cloning specific segments of the mammalian genome: Bacteriophage λ containing mouse globin and surrounding gene sequences. Proc. Natl. Acad. Sci. 74:4406-4410. 1977.
22. Hynes, N.E., Groner, B., Sippel, A.E., Nguyen-Huu, M., and Schutz, G. mRNA complexity and egg white protein mRNA in mature and hormone-withdrawn oviduct. Cell 11:923-932. 1977.
23. Martial, J.A., Baxter, J.D., Goodman, H.M., and Seeburg, P.H. Regulation of growth hormone messenger RNA by thryoid and glucocorticoid hormones. Proc. Natl. Acad. Sci. 74:1816-1820. 1977.
24. Temin, H.M. and Mizutani, S. RNA-dependent DNA polymerase in virions of Rous sarcoma virus. Nature 226:1211-1213. 1970.
25. Baltimore, D. RNA-dependent DNA polymerase in virions of RNA tumor viruses. Nature 226:1209-1211. 1970.

26. Scheller, R.H., Dickerson, R.E., Boyer, H.W., Riggs, A.D., and Itakura, K. Chemical synthesis of restriction enzyme recognition sites useful for cloning. Science 196:177-180. 1977.
27. Sgaramella, V., Van de Sande, J.H., and Khorana, H.G. Studies on polynucleotides, C. A novel joining reaction catalyzed by the T_4-polynucleotide ligase. Proc. Natl. Acad. Sci. 67: 1468-1475. 1970.
28. Villa-Komaroff, K., Efstratiadis, A., Broome, S., Lomedico, P., Tizard, R., Naber, S.P., Chick, W.L., and Gilbert, W. A bacterial clone synthesizing proinsulin. Proc. Natl. Acad. Sci. 75:3727-3731. 1978.
29. Chang, A.C.Y., Nunberg, J.H., Kaufman, R.J., Erlich, H.A., Schimke, R.T., and Cohen, S.N. Phenotypic expression in E. coli of a DNA sequence coding for mouse dihydrofolate reductase. Nature 275:617-624. 1978.
30. Cohen, S.N., Chang, A.C.Y., Boyer, H.W., and Helling, R.B. Construction of biologically functional bacterial plasmids in vivo. Proc. Natl. Acad. Sci. 70:3240-3244. 1973.
31. Bolivar, F., Rodriguez, R.L., Betlach, M.C., and Boyer, H.W. Construction and characterization of new cloning vehicles I. Ampicillin-resistant derivatives of the plasmid pMB9. Gene 2:75-93. 1977.
32. Betlach, M., Hershfield, V., Chow, L., Brown, W., Goodman, H.M., and Boyer, H.W. A restriction endonuclease analysis of the bacterial plasmid controlling the Eco Rl restriction and modification of DNA. Fed. Proc. 35:2037-2043. 1976.
33. Curtiss, III, R., Pereira, D.A., Hsu, J.C., Hull, S.C., Clarke, J.E., Maturin, L.S., Goldschmidt, R., Moody, R., Inoue, M., and Alexander, L. Biologic containment. The subordination of Escherichia coli K-12. In: Recombinant Molecules: Impact on Science and Society (Miles International Symposium Series No. 10). R.F. Beers, Jr. and E.G. Bassett, Eds. Raven Press, NY. pp. 45-46. 1977.
34. Thomas, M., Cameron, J.R., and Davis, R.W. Viable molecular hybrids of bacteriophage lambda and eucaryotic DNA. Proc. Natl. Acad. Sci. 71:4579-4583. 1974.
35. Enquist, L., Tiemeier, D., Leder, P., Weisberg, R., and Sternberg, N. Safer derivatives of bacteriophage λgt·λC for use in cloning of recombinant DNA molecules. Nature 259:596-598. 1976.
36. Tiemeier, D., Enquist, L., and Leder, P. Improved derivative of a phage λ EK2 vector for cloning recombinant DNA. Nature 263: 526-527. 1976.
37. Blattner, F.R., Williams, B.G. Blechl, A.E., Dennison-Thompson, K., Faber, H.E., Furbug, L.A., Grunwald, D.J., Kiefer, D.O., Moore, D.D., Schinn, E.L., and Smithies, O. Charonphages: Safer derivatives of bacteriophage lambda for DNA cloning. Science 196:161-169. 1977.

38. Leder, P., Tiemeier, D., and Enquist, L. EK2 derivatives of bacteriophage lambda useful in the cloning of DNA from higher organisms: The λgtWES system. Science 196:175-177. 1977.
39. Kourilsky, P., Gros, D., Rougeon, F., and Mach, B. Transcription of a mammalian sequence under phage λ control. Nature 267:637-639. 1977.
40. Nussbaum, A.L., Davoli, D., Ganem, D., and Fareed, G.C. Construction and propagation of a defective simian virus 40 genome bearing an operator from bacteriophage λ. Proc. Natl. Acad. Sci. 73:1068-1072.
41. Villarreal, L.P. and Berg, P. Hybridization in situ of SV40 plaques: Detection of recombinant SV40 virus carrying specific sequences of nonviral DNA. Science 196:183-185. 1977.
42. Marians, K. J., Wu, R., Stavinsky, J., Hozumi, T., and Narang, S.A. Cloned synthetic lac operator DNA is biologically active. Nature 263:744-748. 1976.
43. Heyneker, H.L., Shine, J., Goodman, H.M., Boyer, H.W., Rosenberg, J., Dickerson, R.E., Narancy, S.A., Itakura, K., Lin, S., and Riggs, A.D. Synthetic lac operator DNA is functional in vivo. Nature 263:748-752. 1976.
44. Kedes, K.H., Chang, A.C.Y., Houseman, D., and Cohen, S.N. Isolation of histone genes from unfractionated sea urchin DNA by subculture cloning in E. coli. Nature 255:533-538. 1975.
45. Chang, A.C.Y., Lansman, R.A., Clayton, D.A., and Cohen, S.N. Studies of mouse mitochondrial DNA in Escherichia coli: Structure and function of the Eucaryotic-Procaryotic chimeric plasmids. Cell 6:231-244. 1975.
46. Roozen, K.H., Fenwick, Jr., R.G., and Curtiss, III, R. Synthesis of ribonucleic acid and protein in plasmid-containing minicells of Escherichia coli K-12. J. Bacteriol. 107:21-33. 1971.
47. Megaher, R.B., Tait, R.C., Betlach, M., and Boyer, H.W. Protein expression in E. coli minicells by recombinant plasmids. Cell 10:521-536. 1977.
48. Struhl, K., and Davis, R.W. Production of a functional eukaryotic enzyme in Escherichia coli: Cloning and expression of the yeast functional gene for imidazoleglycerolphosphate dehydrase (his 3). Proc. Natl. Acad. Sci. 74:5255-5259. 1977.
49. Clarke, L., and Carbon, J. Functional expression of cloned yeast DNA in Escherischia coli: Specific complementation of arginosuccinate lyase (arg h) mutations. J. Molec. Biol. 120: 517-532. 1978.
50. Mercereau-Puijalon, O., Royla, A., Cami, B., Garapin, A., Krust, A., Gannon, F., and Kourilsky, P.: Synthesis of an ovalbumin-like protein by Escherichia coli K12 harbouring a recombinant plasmid. Nature 275:505-510. 1978.
51. Seeburg, P.H., Shine, J., Martial, J.A., Ivarie, R.D., Morris, J.A., Ullrich, A., Baxter, J.D. and Goodman, H.M. Synthesis of growth hormone by bacteria. Nature 276:795-798. 1978.

52. Rodriguez, R.L., Bolivar, F., Goodman, H. M., Boyer, H.W., and Betlach, M.: Construction and characterization of cloning vehicles. In: Molecular Mechanisms in the Control of Gene Expression. D.P. Nierlich, W.J. Rutter and C. F. Fox, Eds. Academic Press, New York. 1976. pp. 471-477.
53. Sutcliffe, J.G. Nucleotide sequence of the ampicillin resistance gene of Escherichia coli plasmid pBR322. Proc. Natl. Acad. Sci. 75:3737-3741. 1978.
54. Broome, S., and Gilbert, W. Immunological screening method to detect specific translation products. Proc. Natl. Acad. Sci. 75:2746-2749. 1978.

MOLECULAR BIOLOGY AND PUBLIC HEALTH*

Robert R. Wagner, M.D.

Department of Microbiology, School of Medicine
The University of Virginia
Charlottesville, Virginia 22908

Much of my time preparing for this talk was spent trying to decide the meaning of the title. I will be cowardly and deny being its inventor or perpetrator. I hereby blame it on June Osborn, who may also deny it. My version of its history goes back to a telephone conversation with June during which time she lured me into attending this meeting, an invitation I accepted with alacrity. After agreeing to come, she said ever so sweetly: Of course, you will give a talk, an after-dinner talk. Why don't we put you down for a title such as "Molecular Biology and Public Health", which will give you lots of leeway? My audible gasp was taken as enthusiastic agreement. So, here I am, take it or leave it.

Only after receiving the preliminary program did I really understand the mission of this meeting and its Zeitgeist. Therefore, I felt compelled to direct my remarks, at least vaguely, in the direction of cell substrates, a subject of enormous theoretical and practical importance. The problem which confronted me was how to avoid rambling around a subject so broad and diverse. I was encouraged by the subtitled of this meeting, "Their use in the production of vaccines and other biologicals". Knowing virtually nothing about "other biologicals", I could dismiss this aspect summarily. I once knew something about vaccines, much of it antiquated and stemming from the days of Louis Pasteur. My spirits were buoyed. Then I realized that many others, far more versed than I can pretend to be,

---

*Much of this paper was excerpted from my contribution to the virology Task Force of the National Institute of Allergy and Infectious Diseases.

were going to tell you all about viral vaccines. There was I, hoist on my own petard.

Help came from a revelation that recalled to me the penetrating insight of one of the wisest men of our time, André Lwoff. On November 20, 1972, I had the good fortune of attending a "Sesquicentennial Commemorative Symposium" in New Orleans convened by the American Society for Microbiology to pay hommage to Louis Pasteur on the occasion of the 150th anniversary of his birth. The guest of honor and major speaker was André Lwoff. After his usual eloquent speech, Dr. Lwoff was presented the key to the city of New Orleans. We were then shepherded into a bus to be driven to a luncheon to honor Dr. Lwoff and I had the privilege of sitting next to him on the bus. I had the temerity to ask the Nobel Laureate what he was going to do with the key to the city of New Orleans. "Ah", he said, "I shall look for the door."

The wit and wisdom of that reply sank in slowly and then overwhelmed me with its enormity. Truly, there are thousands of keys to the city of New Orleans, as well as Paris and Rome (Urbi et Orbi), but very few can find, much less open any doors. This, I believe, is what Dr. Lwoff was trying to tell me and the rest of the world: We spend our lives (certainly including our scientific lives) in quest of the keys to the kingdom without any real insight into where the kingdom lies or even whether it exists. Perhaps, we would do better looking for doors rather than for keys. Phrased in a different mixed metaphorical context, all of us are devoting our efforts to finding biological and medical answers (the keys) but few of us have the wit or wisdom to devote our energies looking for the question (the door) behind which lies the answer. The research literature is inundated with answers (keys) but contains little in the way of legitimate questions (doors).

The moral of my fable is, perhaps, that the search for cell substrates may be open to wandering in a Minoan labyrinth with a myriad of keys for which there are no doors. Before we set out to seek the Minotaur, the endlessly wandering labyrinth of cells to be used as substrates to produce "vaccines and other biologicals", we should devote some thought and effort to understanding the biology, molecular or otherwise, of the cell.

However, here we encounter the enormous, frightening problem of a gigantic door, which almost fills the universe and a herculean lock in which no currently conceivable key will fit. Herein lies the problem of the cell biologist, confronted with the cell, or an organ with diverse cells whose complexity boggles our minds and dwarfs our technology. What we need is a simple, encompassable cell (a small door) for which we can find a key that an ordinary, mortal locksmith can fabricate. Once we find the small door with the small lock in which we can fit a small key, we can then broaden our horizons and look for larger celestial doors. In other words, start with a minicell to pose the questions, which when answered, permit you to move on to a macrocell (a real cell), thence to an aggregate of interacting cells and a multicellular organism which gradually takes on human proportions.

Nature has possibly, I emphasize possibly, provided us with a spectrum of minicells (or rather microcells) to begin our investigations. These minicells are called viruses, the simplest of which are understood in intimate molecular detail and parts of which can be constructed and assembled in a cell-free system using well characterized chemicals. Some of these molecular aggregates can carry out certain limited functions that sometimes closely resemble the functions of cells (but sometimes do not). Unlike the whole cell, these functions of viruses can be examined in isolation under reasonably well controlled conditions. It is my contention that once we understand the molecular biology, chemistry and physics of these reactions (these small doors) that we can then apply the principles to complex cells into which we can, hopefully, insert keys that fit the locks in the large door.

I had the privilege during the past year of serving on the Virology Task Force of the National Institute of Allergy and Infectious Diseases. The mission of Panel V of the Virology Task Force was to explore the prospects for research on viruses to be used as probes for studying intricate processes of complex cells, some of which could be used as substrates. It seems to me that the mission of Panel V of the Virology Task Force carries certain important implications for this meeting at Lake Placid on "Cell Substrates and Their Use in the Production of Vaccines and Other Biologicals". Therefore, I have taken the liberty of quoting, almost verbatim from my own contribution to Panel V of the Virology Task Force.

The major thrust of the deliberations and report of Panel V is that viruses provide ideal model systems for uncovering the fundamental principles underlying the molecular organization, genetics and biological activities of cells and, ultimately, multicellular organisms, including humans. The huge gaps in our knowledge and the many misconceptions about diseases are due in large measure to directing our research efforts toward inordinately complex and uncontrollable systems. Nature provides very few models that enable scientists to isolate the phenomena they wish to study; therefore, multifactorial systems often give misleading information because the investigator cannot devise logical experiments that provide simple yes or no answers. The simplicity of viruses provides us with an unique opportunity to use them as probes for studying isolated phenomena that often mimic the biological and chemical events in the host cell. It is our contention that virology, as a discipline, has progressed to the stage where it can be applied to in-depth investigation of the molecular events which determine cellular processes as they relate to health and disease.

To understand the rationale behind our proposal for promoting future research in virology as applied to cell biology and pathology, it is essential to recall the history of the science of molecular biology that provides the fundamental basis for our understanding of biological function and disfunction. The science of molecular biology began when a small cadré of scientists, more versed in physical sciences than in classical biology, agreed in the 1940's to pool their intellectual and physical resources (their software and their hardware) in studying the molecular structure, organization and function of an encompassable biological system. The most crucial decision was to choose a system. With great foresight, their decision was based on the simplicity of the organism, the feasibility of producing large quantities suitable for chemical analysis, ability to attain purity and homogeneity, and, to some extent, the capacity to make the observations in minutes to hours rather than in days to years. Delbrück and his colleagues chose the bacteriophage, a fairly simple virus of bacteria, that satisfied many of the conditions essential for their research. (Actually, the one they chose, inadvertently turned out to be the most complex of them all.) This decision paid off in enormous research dividends, resulting in a revolution in the science of biology which provides a means for asking legitimate questions of all biological systems. These efforts also resulted in quite a large number of Nobel Prizes.

The application of these discoveries in the molecular biology of bacterial viruses led to the methodology and concepts that provided the tools and stimulus for investigation of the bacterial cells which are the hosts for infection by these viruses. By these means, the genetics, biochemistry and physiology of bacteria have progressed enormously. Viruses provided the model systems for these studies, many of which have resulted in major advances in our knowledge of bacterial diseases such as cholera and diphtheria.

Another large dividend emanating from this research on the molecular biology of bacterial viruses was the ability to devise, at long last, a scientific basis for studying viruses that infect animals and man. The very recent science of molecular animal virology is a direct offshoot of the fundamental science of bacterial virology. The explosion in animal virus research is exemplified by the estimate made by N. P. Salzman (personal communication) that 90% of all the published work in this field has appeared during the past ten years. Equally impressive might be an estimate of the number of biochemists and biophysicists who have been attracted to the field of animal virus research. Armed with the fundamental principles established by the molecular bacterial virologists, the molecular biology of animal viruses has advanced to a stage where considerable information is now available to define the fundamental molecular structure and function of most animal viruses. Much remains to be done, of course, but the day is perhaps not too far distant when most of the major problems in molecular virology will have been solved.

The challenge facing the scientific and lay communities is how to apply the plethora of knowledge still pouring out of the elegant but somewhat esoteric research in molecular virology. Of highest priority, of course, is the application of this fundamental knowledge to uncover means for eliminating viral diseases. Single component, safe vaccines, produced by synthesis of viral proteins in test tubes, are now technically feasible. Chemotherapy of viral infections poses a far more formidable challenge but the techniques are probably available at least to screen antiviral agents.

The next major challenge for molecular virologists is to apply their knowledge to solving the thorny problems of cell biology, particularly as they apply to medicine. It seems reasonable to expect that this research in molecular cell biology of isolated systems is the next great frontier. The time has come to exploit this truly fundamental knowledge of simple virus systems to the field of cellular biochemistry, physiology, genetics and pathology. This would seem to be the next logical step in the steady progression of research aims in biomedical science from the lowly bacterial virus to man. It should be kept clearly in mind that success of such a venture requires a deliberate and logical series of steps; it would be wasteful of scarce resources to skip past the basic field of molecular biology for a tempting but foolhardy direct assault on poorly understood human diseases. The lesson of biomedical history seems clear: that promised rewards of direct applications of incomplete, and often fallacious, data inevitably results in dismal failures, wasted resources and disillusion of a once-patient public. The take-home lesson of recent history is that there are no shortcuts and that the next step must, of necessity, be an in-depth basic study of cell biology, to a considerable extent based on experience of the molecular biologist, biochemist and biophysicist.

With these basic promises in mind, Panel V of the Virology Task Force proposed a series of recommendations for programs of future research in the field of molecular cell biology using viruses as probes to uncover underlying cell principles. A few examples of prospective research are summarized below.

1. PROTEIN SYNTHESIS

Viruses represent powerful tools for studying the role of protein synthesis in the economy of a cell and the way in which specific products can determine the functions for which each cell is designed.

Mechanisms are available for the cell to call forth synthesis of specific proteins designed to perform specific functions. Moreover, protein can be modified after their synthesis in ways which promote more efficient functioning or for assembly into essential cell structures. These properties are much easier to study in viruses because of their relative simplicity. Viral proteins can perform similar functions which are much more readily studied because of their limited number. We recommend support of research in the use of viruses to study a wide variety of translational controls of protein synthesis and processing that are also important for understanding more complex cellular functions:

i. Kinetics of protein synthesis should be studied both inside infected cells and in cell-free systems by the use of specific viral messengers, many of which are well characterized.

ii. Regulation of translation by means of specific proteins reacting with messenger RNA can be studied in greater depth by use of viral gene products, such as the gene 32 product of T4 phage or other viral products. Studies such as these should divulge basic principles for feed-back regulations for controlled synthesis of specific cellular proteins. The means by which certain proteins can react with specific chromosomal sites to call forth or inhibit protein synthesis can be far more readily analyzed by using virus nucleic acids as model systems.

iii. The capacity of viral messenger RNA to compete for ribosomes provides an important model system for translational controls of cells which require specific proteins for their specialized functions. This capacity of viral messengers to subvert cell protein synthesis should be studied in depth in order to learn what controls synthesis of specific cell proteins.

iv. Post-translational cleavage of viral proteins provides the ideal system for probing similar cleavage of cell proteins to make functional components serve the requirements of the cell and distant tissues.

v. The utility of specific cell proteases in controlling cell functions can be studied more readily by altering these functions by viral infection and/or transformation.

vi. Many different secondary modifications of cell proteins can be investigated far more successfully by using viruses as tools to study specific alterations of viral proteins by means such as glycosylation, methylation, phosphorylation, sulfation, etc.

vii. Subcellular localization of proteins should be first studied by using specific, readily identifiable viral proteins.

viii. Membrane translocation, transport and secretion of proteins can be and should be studied in virus-infected cells because of greater ease in designing experiments with viruses to explore cellular mechanisms for controlling these essential cellular functions.

ix. The extremely important mechanism by which proteins self-assemble into organelles can be first investigated by using viruses as model systems for studying the physical properties of macromolecules which regulate their maturation into functional structures essential for the economy of the cell.

2. MEMBRANE COMPOSITION, ORGANIZATION AND FUNCTION

Enveloped viruses (such as influenza, paramyxoviruses, rhabdoviruses, herpesviruses and oncoviruses) provide ideal models for studying lipid assembly and membrane organization in a well defined biological system. It is recommended that research efforts be directed toward the following studies: a) lipid composition of viral membranes compared with membranes of the host cell; b) physical properties and dynamics of viral membranes as studied by X-ray and electron diffraction, nuclear magnetic resonance and electron spin resonance spectroscopy, fluorescence depolarization and differential calorimetry; c) lipid asymmetry as studied by labeling procedures and susceptibility to lipases and proteases; d) surface reactivity to lectins and specific antibodies; and 3) reactivity with polyene antibiotics. All these investigations should shed considerable light on structure and assembly, as well as possible functions of the far more complicated membranes of eukaryotic cells, as briefly outlined below:

i. Lipid exchange in viral membranes reveals that enveloped viruses may provide ideal systems for studying lipid exchange in interacting cell membrane systems. Virus membrane lipids can be removed, replaced, altered and mingled with specific fatty acid and phospholipid headgroup probes for fine structure studies of membranes. Research efforts can then be directed to study mechanisms of lipid exchange between two membranes and within the two layers of a single membrane (flip-flop) by a) interaction with vesicle bilayers of known lipid composition, b) cholesterol depletion from and repletion of viral membranes, and c) phospholipid exchange between viral and other membranes by means of the phospholipid exchange proteins found in various tissues.

ii. Models for studying de novo membrane biosynthesis are provided by poxviruses and the Pseudomonas bacteriophages, which promise by intensive study to provide essential data on mechanisms which regulate membrane biosynthesis. These two groups of non-budding viruses are quite different from budding viruses which only rearrange preformed lipids of the host cell membrane for their own membrane organization.

iii. Membrane insertion of proteins can probably best be studied by using enveloped viruses as model systems. Available techniques provide incisive probes for following the progression of newly synthesized membrane proteins from membrane insertion to budding. We recommend development of research programs using viruses as model systems to study the following critical questions: a) glycosylation of membrane proteins, b) their transport from endoplasmic reticulum, Golgi apparatus, and plasma membrane, c) insertion into lipid vesicles and vesicle fusion, and d) secretion of glycoproteins to the environment.

iv. Glycolipid synthesis and membrane insertion can also be studied with potentially great profit by using enveloped viruses as model systems. We recommend that research programs be mounted to exploit viral membranes to study: a) selective insertion of glycolipids at sites of virus budding, b) glycolipid turnover during viral infection and glycoplipid turnover.

v. The membrane matrix (M) proteins of myxoviruses, paramyxoviruses and rhabdoviruses provide very promising tools for investigating analogous proteins of eukaryotic cells such as erythrocyte spectrin and actin. Important advances can be made in knowledge of membrane assembly and function by supporting the following research: a) membrane location and insertion of M protein; b) association of matrix protein with glycoprotein trails; c) physical relationship and proximity of matrix protein to nucleocapsid protein, d) membrane protein interactions studied by use of reversible bifunctional protein crosslinkers; and 3) role of matrix protein in membrane fluidity.

vi. Peripheral membrane proteins of host cell origin should be studied in viral membranes, which could provide clean, homogeneous and readily controlled model systems for investigating these important components of all mammalian cells. Of particular interest for intensive study would be: a) histocompatibility antigens, b) lectin binding sites, and c) a considerable variety of host cell enzymes, such as proteases, glycosidases and lipases.

Many other examples could be cited to emphasize how viruses can be used as incisive probes to study cell biology. It is my contention that the scientific establishment must exploit the basic systems that nature provides as simple models for understanding the generality of biological systems. We must convince the lay public that fundamental research at the molecular level is likely to be the avenue that leads to the solution of the medical and environmental problems that beset us. Not only does this approach promise more reliable information but, in the long run, will be far cheaper than emotional crusades. We should remember that the misguided crusaders seeking the Holy Grail almost destroyed Christianity and western civilization.

SUMMARY AND CONSIDERATIONS FOR THE FUTURE

Introductory Remarks

Thomas C. Merigan, M.D.

Division of Infectious Diseases
Stanford University Medical Center
Stanford, CA 94305

It is my pleasure to introduce the final session of our meeting on cell substrates. So far at this meeting we have discussed the requirements for cells as substrates and the philosophy regarding choice of primary versus continuous cell lines. In addition, we have discussed the use of such cells in the production of newer, live viral vaccines and future needs in regard to the development of new inactivated vaccines and biologicals such as interferon and urokinase. It seems likely that molecular biology will provide us with effective tools to slay the DNA "dragon" which Dr. Hilleman correctly points out represents the genetically dangerous principle in cell lines. In my view, however, it is still somewhat in the future for recombinant DNA to provide large quantities of biologically important molecules for human medicine. Actually the real question with any new (or old) approach is the risk-benefit issue or, to paraphrase an idea of many years ago, "Do the ends justify the means?" Drs. Dull and Cook in this session will provide us with information about the public and its perceptions and needs for involvement in these issues today, as well as the difficult but very important ethical issues surrounding vaccine development and research. Although the subject matter of this meeting has ranged widely, those who have convened it must be congratulated in that their goal of examining critical issues in a significant area of public health medicine for a scientific consensus has been met. All of us will leave this meeting with a deeper understanding of the issues in regard to new vaccine development and with a greater awareness of each other's positions.

BENEFIT-RISK ASSESSMENT

H. Bruce Dull, M.D., Assistant Director for Program

Center for Disease Control, Public Health Service

Atlantic, GA

Those who organized the program for this symposium have, indeed, done a skillful job. Scientific presentations on previous days focused attention on exciting new technologies and their obvious utilities. Speakers and discussants led us onto a mountain of expectation. They portrayed the possibilities and probabilities of cell substrates capable of making vaccines and other biologicals more precise and of higher quality than would otherwise be possible.

Today's agenda seems designed to bring us from that mountain - or if not all the way down, at least to an altitude where we can see our starting place. I am called on to do this "reality testing," if you will, by describing the concept of balancing the easy-to-grasp benefits of the new technologies with the sure-to-be-found risks. Dr. Cook, who follows me, is asked to review another aspect of today's realities by considering the ethical implications of new technologies when they result in vaccines for humans.

I asked myself, as maybe you did, why this particular program strategy was chosen - why the implied cautions were not positioned earlier on the agenda, or why they were included at all. Do scientific/technical meetings need such admonitions? I'm sure you have all noted that increasingly commonly, scientific meetings, particularly those organized in the last few years, consciously schedule "balancing" presentations, those that attempt to indicate that scientists' enthusiasm for scientific advancement must acknowledge the public's interest in being involved and knowing what risks the new technology may carry with it. Part of the apparent need for today's presentations, as I thought about them, is a bit like the American television commercial showing a man slapping his face with after-shave lotion who says, "Thanks, I needed that."

In fact, the reason for this line of programming is obvious. It is to encourage that we anticipate rather than later be confronted by problems that could reasonably be expected to accompany scientific advances. And it is to recognize that a good analytical procedure for doing this is one that consciously balances the human values and the human perceptions with the technical achievements and scientific probabilities.

THE PUBLIC AND ITS PERCEPTIONS

Risk, in a general sense, has reached a new level of importance in determining public policy. Likewise, human values, which always have been important in biomedical research, have become far more dominant in post-World War II and particularly in the last decade. There are three major reasons: First, the public has begun to exert more and more direct control on the world and its institutions. Control is felt in myriad ways ranging from legislation, to funding, to court actions. In research, control has grown out of what some people describe as a "love-hate relationship" in which the public and the research enterprise are simultaneously in harmony and at odds. On the one hand, the public has an unbounded expectation for what research, particularly biomedical research, can achieve, and on the other, it has deep concern over things like experimentation involving human subjects or the potentially undesirable results of uncontrolled recombinant DNA technology. A list of tensions between science and the public on what-can-be-achieved and what-is-sure-to-go-wrong could be as long as one would care to make it.

Second, the public with a new assertiveness is seeking for meaningful ways to participate directly and personally in decision making when public interests and public resources are involved. In that biomedical research is largely supported in the United States and in many other countries with public funds, public leverage has been increasingly felt in determining both research objectives and their relative importance.

Third, and probably the most difficult to deal with, today's public expresses a philosophy of its "entitlement" - not just expectation, but entitlement - to the goods and services society can potentially offer. What I speak of is not only entitlement to information and being involved but also entitlement to an essentially risk-free environment or, at least, to some sort of stand-by management of unexpected, adverse events that interrupt what is perceived to be just. I don't want to dwell on the philosophy of public entitlement other than to indicate that it has become such a dominant characteristic of contemporary society that any assessment of benefits and risks must recognize the public's expectation of involvement and control.

If I may digress briefly at this point, I would like to refer you to a 1976 conference, one held in Geneva and co-sponsored by the World Health Organization, the World Medical Association, the Committee on International Organizations of the Medical Sciences, the International Association of Biological Standarization, and the United States Public Health Service. It was titled the "Role of the Individual and the Community in Biological Research, Development, and Use," and, I believe addressed the kinds of issues that are at the center of our discussion. The Proceedings of the conference were published in 1977 as a Supplement to the Bulletin of the World Health Organization (Supp. 2, Vol. 55, Bulletin WHO, 1977).

MINIMIZING LIABILITY

Besides an understandable personal identification with health and disease, there is another reason why the public's attention has focused on biomedical research. This is because of the courts. In hearing complaints of personal injury associated with medical treatment or with vaccines, courts have applied various principles of liability, making the issue of liability of first order importance for people involved in vaccine research, production, and utilization. Without analyzing the liability issues in detail, suffice it to say that minimizing liability associated with the inherent risks in vaccines has been one of the most controlling issues in planning or conducting vaccination programs in recent years. Not only have legal concepts of negligence in vaccine manufacture and administration been replaced by theories of absolute liability - including the duty to warn recipients of known or likely risks - but also the climate created by concerns over liability has engulfed vaccine research and production and the policy-making and implementation of public health vaccination programs. The Swine Influenza Program of 1976, for example, was stalemated and could not proceed much beyond its earliest stages until Federal legislation was enacted to manage potential public risks.

Since one of my assumed responsibilities in these remarks is not only to discuss the assessment of risks possibly associated with new substrates for vaccine production but also to suggest how they might be minimized, let me offer suggestions: The first is to be able clearly to demonstrate the need or ultimate usefulness for new or modified vaccines. At the risk of seeming to say the "old is the best," developing new and, particularly, alternative vaccines these days requires a persuasive demonstration of need. This is obviously because newness is perceived as being synonymous with unknown or untried - with yet-to-be recognized risks. Substitute or updated vaccines that replace those with known dimensions of utility and risk are even more vulnerable to public apprehension because they are without a history to compare with what is now in place.

The second way to minimize perceived risk is to assure that new vaccines will work effectively and be efficient. This is not only in terms of their economics but also their acceptability as seen by the public.

A third and obvious one is to assure the highest possible degree of vaccine safety for public use by presenting evidence that is as clear and well documented as possible. In addition to the scientific evidence of safety, there must be indication of impeccable clinical trials which can withstand challenges to their science and their humanity. There must be sufficient consistency in findings to allay the anxiety produced when an aberrant result is mistaken for a valid observation. When this does occur, efforts to explain or "neutralize" it by carefully designed, promptly executed supplementary studies become exceedingly important.

Fourth, and readily apparent to an international conference like this, is to "reinforce" the results of research, development, and production in consistent ways at different times in different places. Collaboration, cooperation, and communication among independent groups at all points in the evolution of vaccines gives added assurance that reinforcing results will occur. To this end, protocals for clinical studies should be "standardized" so that observations can be consolidated. The internationality of vaccine research and development best serves the public interest when it demonstrates scientific consistency and permits rapid assembly of evidence on need, utility, and safety.

The fifth means of minimizing liability - and probably the most important one of all - is to involve the public at all suitable points in the planning, implementation, and evaluation of new or alternative vaccines. Whether this calls for a determination of need or an evaluation of utility, the public interest as an important objective of vaccine research and a controlling influence must be recognized. Unfortunately, we do not yet have good procedures with which to capitalize on the public's assertiveness and interest in being involved; but recognizing the public's determining role in enterprises like vaccine research, development and use, efforts to constructively utilize public interest should carry a high priority.

BENEFIT-RISK ASSESSMENT

Let me turn now to examining the benefit-risk assessment as a useful procedure in planning and evaluation, particularly as it might apply to vaccines. At the outset, I should indicate that to me the process of systematically balancing benefits and risks has more inherent value as a management tool than as a scientific or mathematical analysis. This is so, because, first of all, judgment which determines which factors are going to be balanced - either the benefits or the risks - is highly subjective. If we talk about public benefits and individual benefits or if we try to quantify things like health or life or death, we are dealing with personal and hardly objective matters. While we may be abe to quantify categories like vaccine research and development, production, and distribution or like medical care, hospitalization, and such by measuring effort or cost, it is nearly impossible to assign an objective value to preventing an epidemic or assuring freedom from an infectious disease or enjoying health and well-being. To make quantification an even more unrealistic exercise is to recognize the variation in values from country to country where priorities in health and research are bound to vary greatly and where the tolerance for risks and the interpretation of benefits will have no uniform pattern whatsoever.

But even admitting its shortcomings, the process of examining benefits and risks is still very useful. This is because the process requires one to isolate viewpoints - scientific and public - which focus on important comparatives. Benefit-risk assessment is best considered to be a discipline with a range of useful applications. It can be of particular value in looking at alternatives such as, for example, a new vaccine for an old one, a vaccine produced by method A in contrast to a vaccine produced by method B. Here quantifiable differences may still be difficult to portray with precision, but the qualitative differences and the extent of the differences can be highlighted.

RISK-RISK ASSESSMENT

Another analysis in the same "family" of procedures, if you will, is risk-risk assessment. In contrast to the benefit-risk comparison which largely examines contemporaneous elements, the risk-risk comparison often introduces the perspective of time. Let me try to explain by using a hypothetical illustration. Consider a vaccine used with some degree of success in preventing an infectious disease. At the time of our benefit-risk assessment intended for use in program planning, cases of the disease continue to occur even though a great deal of vaccine is being used on a regular basis. The disease in question has definable severity and complications, and the medical needs with respect to diagnosis and management are resonably well known. The vaccine being used has an established level of effectiveness and a known small risk of characteristic adverse reactions. The question at hand concerns the soundness of investing additional resources to bring the disease under optimal control.

A benefit-risk analysis of the question selects a series of benefits reasonably expected with program intensification, attempting to quantify them in appropriate ways and compare that more salubrious time in the future with associated risks (and costs) to give useful contrasts. What risks should be examined? Obviously they include risks from the intervention itself not only in terms of the costs of the intervention - in this case vaccine and its delivery - but also the risks of any adverse results which could be expected. Although it is possible to select items for balancing that have some degree of equivalency, the particular items chosen can obviously influence the outcome. Some people claim that artful choices come close to manipulating.

For me, a risk-risk analysis of the same fact situation is easier to portray. In this case, the risks chosen for interpretation should reasonably be the risks of the disease as it was occurring in the past, is occurring at present, and may be occurring in the future. The quantification of these risks is relatively straightforward once the particular qualitative and quantitative dimensions are selected. The risks (and costs) of the interventions are of the same sort as in the benefit-risk assessment but can more easily be made to parallel the risks of disease. In simplest terms, a benefit-risk analysis examines the benefits of not having disease with the risks of doing something about it. The risk-risk comparison examines the risks of disease and the risks of disease control. Both obviously can be helpful and ultimately can be used where they are best suited.

SUMMARY AND CONCLUSIONS

In summary, it is quite clear that this meeting has provided an opportunity to describe exciting areas of innovation in cell substrate technology as it applies to vaccines. Much of it has obviously practical application in improving the research and development of vaccines in terms of yield, purity, precision, cost, safety, and applicability. Although many of these outcomes can easily be understood,others - especially those to do with safety - are likely to raise concerns when viewed by the public and, to a degree, the health professions. There are public perceptions about scientific innovations generally and special concerns about those which relate to biomedicine and human health. In that the public has been sensitized to questions of genetic manipulation and cancer etiology, these particular aspects of the cell substrate debate are particularly vulnerable to misunderstanding. The public obviously needs to be regularly informed about progress and about potential benefits; and to the degree that benefit-risk or risk-risk analyses can sharpen some of the issues, they can be useful.

In noting that this meeting has carried participants onto a mountain of expectation, it is reasonable to recommend that the public now be brought along to experience a comparable exhilaration. The public needs to share in the excitement over technical advances and to help identify and deal with the risks as well as the benefits that technical advancement inevitably brings.

ETHICAL CONSIDERATIONS IN THE SELECTION OF CELLS AND THE USE OF BIOLOGICALS

Robert E. Cooke, M.D.

The Medical College of Pennsylvania, 3300 Henry Avenue, Philadelphia, Pennsylvania

Dr. Bruce Dull has discussed in-depth the benefit risk assessment in the development of and usage of immunizing agents. It is my task to cover the wide range of other ethical considerations which face investigators and makers of public policy concerned with immunization. Many of the issues have been explored in-depth at a National Conference on Immunization sponsored by HEW in November 1976, and by a series of National Immunization Work Groups convened by the Assistant Secretary for Health of Health, Education and Welfare thereafter.

"There seems to be general consensus that the development and testing of new vaccines on human beings will pose increasingly complex problems for investigators and for the public authorities which must review and approve prelicensure studies." (1).

Litigation, court action and large damage awards for vaccine associated adverse reactions, often without fault or negligence in production or application have been added to more traditional ethical issues of consent, selection of subjects as well as risk-benefit analysis.

In a discussion of vaccine development it is superfluous to emphasize the importance of immunization. Next to improved sanitation there has not been a greater contribution to world health. Likewise, it is not necessary to dwell on the fact that sooner or later regardless of new methods of development, animal or *in vitro* testing trials on humans will be required. Further it is generally agreed that it is essential that these trials be limited to as small a number as possible initially to gain specific and reliable answers regarding safety and immunogenicity in those least likely to suffer serious and irreversible adverse effects prior to widespread general usage. Therefore it is inevitable that some persons must be subjected to greater risks with less likelihood of benefit than others.

With the above as given it is worth reaffirming that the basic ethical principles underlying research which are in general equally relevant to the application of that research are beneficence (including the prevention of harm as well as doing good), respect for persons and justice.

In the discussion of benefit risk assessment beneficence and the prevention of harm are obviously the major considerations. My discussion therefore will be based primarily on the practical ethical issues which arise from the application of the other two principles.

Respect for persons is concerned with the appropriate recognition of the freedom and autonomy of the individual. It acknowledges that human beings - i.e., mature persons - are capable of making important decisions concerning their own lives without undue influence, coercion or indoctrination. Autonomy can be thought of as the ability to act freely and responsibly; as the ability to consider consequences prior to action even though the action may not be in keeping with the accepted norm.

The autonomy of the individual has been recognized traditionally by the process of informed consent. The Nuremburg Code, the Helsinki Declaration and other codes basically depend upon consent to assume that subjects of human experimentation are indeed treated as humans.

The requirement for consent in more legalistic terms expresses the principle that a person's physical integrity should not be impinged upon without their informed consent.

Unfortunately as Charles Fried has pointed out (2) that "even if you make every effort to inform people fully, most people will choose without giving sufficient thought to that information." To illustrate - "Doc - I don't care what you do just get this darned sore throat cured." As a consequence of the lack of confidence in consent as a process efforts have been made to make disclosure more and more complete and more and more comprehensible to the point of frightening any and all possible participants or to do away with consent altogether. Case law is filled with examples of judgments against physicians but rarely investigators for inadequately informing patients but there is also a classic suit because of excessive information which deterred a patient from receiving necessary treatment.

Fried proposes and I agree that it is worth living with such imperfections as long as you give people a reasonable opportunity to choose.

The problem arises in vaccine development prior to more widespread trials when risks might be high and benefits low to an individual subject who volunteers. For that matter historically investigators have commonly placed themselves in such a position. Under these circumstances does society have a responsibility to step in and protect the individual from his own volunteering?

In this regard the autonomy of the individual can be thought of as being always in potential if not actual conflict with the forces in society that work to protect the best interests of all individuals. Although paternalism may have a pejorative connotation to many it is an expression of the desire of society to do good and to prevent individuals from doing harm to themselves.

Sharp differences of opinion exist as to whether or not investigators should be permitted to offer to adults opportunities to participate in research where risks of serious and irreversible injury are high. Some argue that if the risks are fully appreciated by the subject and the consent is truly responsible and informed the investigator, out of his respect for the autonomy of the individual, should offer the opportunity to participate. Others argue that even if consent is fully informed, free of coercion and responsible, the investigator has the responsibility to himself and others not to do harm. Autonomy vs. paternalism - both in the best sense of the word.

Even the investigator who serves as his own volunteer subject in vaccine development puts his family at risk from his injury or death. In those circumstances is permission from the family an appropriate requirement?

How far responsibility for others extends is not readily established. All would agree that the investigator and the developer of cell lines and biologicals have the responsibility to minimize risk of harm by animal trials and laboratory approaches, but at what point humans are used cannot be easily determined. When humans are used the risk of harm should be kept as low as possible by using the "phase method" - namely controlled incremental increases in the numbers of human subjects used based on critical evaluation of data at each phase. In this way no extra human beings are put at risk.

It is certainly clear that the responsibility of the investigator extends to the selection of initial subjects at least who are experienced in choosing. The opportunity to participate in the development of biologicals and in early trials should be limited to persons who are used to choosing - not to those whose first opportunity to choose is choosing to be a research subject. Out of respect for the autonomy of persons, subjects should be selected who are free to make decisions on their own and who make routinely many choices in their daily lives.

Persons living in restricted settings such as prisoners, or patients in institutions or those who are inexperienced or limited in their choices by illness, limited capacity, limited emotional stability or immaturity are not appropriate subjects in contrast with free living decision makers. Such a determination does not categorically exclude such persons from participation as research subjects or in trials but very good reasons must be mounted to justify their inclusion.

In the field of vaccine development and vaccine trials special reasons may exist why certain persons with limited decision making experience should be selected. In the testing of the effectiveness of biologicals individuals without prior immunological experience for that agent maybe required. Most adults may not be suitable subjects because of naturally acquired immunity. Likewise adverse reactions to biologicals maybe substantially different at different ages. The febrile reactions to influenza vaccine are an example. Consequently the dangers to the general population of young children cannot be predicted from trials on adults.

Some minimizing of the use of children in the first instance might be possible through the intensive search for consenting adults who were immunologically inexperienced by virtue of isolation or other causes. However, the frequency of adverse reactions in children cannot be obtained under any circumstances from the study of adults.

The use of children, if deemed necessary, presents serious problems. In the development of guidelines for the conduct of research the National Commission for the Protection of Human Subjects of Biomedical and Behavioral Research labored longer and had more difficulty regarding children than it did with any of its other charges. The specific area of vaccine development although not addressed in-depth was cited repeatedly as one of the most controversial.

Hauerwas in his paper to the Commission on Ethical Issues in the Use of Human Subjects affirms this belief. (3). "The ethical issues raised by the use of prisoners and the poor seem simple when compared to the problems involved in the use of children and other non-competents. In order to develop certain kinds of drugs or procedures we can do all the animal and adult testing we want and still we must finally test on children - i.e., a test group who by definition cannot give informed consent. Paul Ramsey has argued that no one, parent or guardian, even with the best intentions has the moral status to consent for a child to be made subject of medical investigation solely for the accumulation of knowledge (except when epidemic conditions prevail). To quote: 'Where there is no possible relation to the child's recovery; a child is not to be made a mere object in medical experimentation for the good to come.' If it is objected that this severely restricts possible advances in childhood medicine, Ramsey argues that the moral progress of the race is more important than the scientific. Thus, testing of children is the paradigm instance that at times it may be necessary to choose between morality and knowledge even though we normally assume that we do not have to choose between them."

When vaccine research or vaccine trials require participation of children their participation could be justified by some on the grounds that benefit to the subjects would result from the immunization and could be considered therapeutic. Parents traditionally have been expected to act in the best interests of their children and therefore could enroll them in a trial that would lead to individual benefit.

On the other hand, it can be argued that the subjects receive no more benefit than other children at a subsequent time and are subjected to far more in the way of risks as initial subjects. Indeed at certain stages of vaccine development it is quite possible that initial subjects would be subjected to risk with only a remote possibility of being protected by a vaccine early in its development. Such subjects might be protected later if the vaccine was perfected but again one group of children had to be subjected to risks in excess of others.

How then can actions be taken that may not always be in the best interests of an individual child subject? Ramsey would argue that such use of a child as a means rather than an end in itself is never justified. No one, he maintains, including the parent has the moral right to volunteer another for someone else's benefit. McCormick condones such participation on the grounds that a child can be presumed to wish to participate in such research since natural law doctrine indicates that individuals owe something to other members of the human species. If one then makes a presumption of what a child would wish, the child would not only wish to do this, but he ought to do this.

Unfortunately I cannot agree with Ramsey or McCormick. The latter's arguments lead to the conclusion that all adults ought to wish to participate in research and so any adult could be drafted as a "volunteer" subject for research. No one is recommending the drafting of adults as research subjects; therefore no one should draft infants or children. That is simple justice.

As regards Ramsey's arguments the limited autonomy of the child and the natural autonomy of the family unit must be appreciated. To pursue this approach it is necessary to analyze in-depth what consent represents. Why is consent important from an ethical perspective? Most people would agree now that even though no risk was involved, no harm, that except for observational activities in public places, consent of someone is necessary if you are doing something with that individual. That means that consent is important for more than the protection of the individual from physical, mental, social or economic harm. Further, most people would agree that consent might possibly provide some protection, but from the writings of Katz it is clear that such protection is limited. Persons are frequently coerced unknowingly. For example, a patient who has a 100 percent fatal disease cannot easily resist the trial of a new therapeutic measure. He is going to die without it - and might live with it. There is obviously not very much freedom of choice in that particular situation. Thus there is a great deal of subtle coercion. Even more subtle than that is the physician relationship with the patient. The patient respects the doctor, the patient is somewhat indebted to the doctor. So it is very difficult, I think for consent to be free of some kind of coercive influences.

If one reads consent forms carefully, I think you would have to agree that in the cancer chemotherapy area, for example, one would have to be a good clinical pharmacologist to understand what is meant. Indeed the Institutional Review Board preliminary report to our Commission indicated that the reading level of most consent forms is about third or fourth year college level, not that of subjects. The comprehensibility was at a very high academic level, and the comprehensiveness was at a relatively low level. Thus in the consent process, as Freeman points out, people can give responsible but uninformed consent. Consent is poor protection. Institutional Review Boards can provide far better protection against physical or mental harm than consent.

Yet consent is said to be important for two other reasons. Respect for the autonomy of the individual and respect for the dignity of the individual. Even though there is no risk, we expect consent out of respect for the freedom of the individual to choose, and out of respect for the person as an individual. How then do these aspects of consent apply to the infant or young child?

Autonomy means freedom to choose. Ability to determine one's own course of life. However, such a concept is meaningless when applied to the infant. The infant does not have any autonomy. He cannot even choose what he wants to eat. He has to be changed when somebody else decides. He is totally dependent, and for a number of years he is relatively limited in terms of his freedom of choice. The notion of autonomy applied to the infant is a meaningless one.

Children can learn to drive at a certain age but we don't allow it at a younger age. We don't allow a child to look at "wicked" movies until he is 16, but at 12 he can look at Peter Pan. There are graded freedoms to choose. We have graded autonomy - a graded freedom to choose.

To return to the infant who has no autonomy, then consent to recognize autonomy is meaningless. Consent out of respect for the dignity of something or someone, does not have to be given by the individual. For example, we don't allow cadavers to be mutilated. Before one can do an autopsy, permission from a caring person must be obtained. We have certain respect for dignity also. We don't allow people to desecrate a mountain. We require some acceptance by the community. The consent to recognize dignity does not have to be given by the mountain. It can be given by some other individual in a proxy situation out of respect for individuals or things. The experiments that were rather horrifying to the Commission in which the dead fetus was decapitated and perfused to study brain metabolism lacked respect for the dignity of the human fetus even though people might have agreed to abortion. Decapitation of a fetal dog and perfusion would probably not generate that kind a revulsion. Consent then can be given to recognize the dignity of other individuals.

When a child is viewed not as an autonomous individual but rather as a dependent part of a family there may be a substantial difference. Trust, love and care for the child is an integral part of family. The obtaining of consent by the family for actions toward that child is out of respect for these bonds not out of respect for the child as an autonomous being. Proxy consent which Bartholome has interpreted to be primarily an attempt to protect the child can equally be seen as an attempt to protect the integrity of the family unit.

It is true that parents may not always know what is best for their children. The important issue is that the historical tradition of family expects that the family should. "In other words proxy consent (or permission) as an institution (or procedure) is one way to insure that whatever is done to the child is done in accordance with the moral conventions and traditions of that family."

Parents as leaders in the family are allowed to make important decisions for the family. Enrolling in research might well be such a decision. In some areas of research the whole family might well participate with the parents giving consent on behalf of the whole family. If the procedures were uncomfortable or considered risky they could remove the child and themselves from participation.

Unfortunately in vaccine development or vaccine trials withdrawal from the experiment after receiving the biological is a meaningless gesture. The event that may produce harm has already occurred.

If parents are permitted to make decisions regarding their children as part of the family's desire to assist others and serious risk of harm is not present is this not in keeping with usual parental responsibility? Should parents be prohibited from making such decisions or should they be encouraged to make responsible decisions and seek out all the consequences before permission is given?

If children must be utilized in vaccine development and vaccine trials, certain necessary steps should be taken. Adequate trials in animals should be carried out including those in immature animals. Trials in adults and emancipated minors and older children who do not object should precede the use of younger children. Children from stable homes in which both parents agree on participation should be used, not orphans or foster children or institutionalized subjects even though more convenient.

Although some persons have argued that a lottery is the only fair approach, the volunteering by families permits the exercise of family autonomy and avoids the police state aspects of forcing families to participate against their wishes.

The principle of justice as well as that of respect for persons has additional ramifications that pertain to the selection of children as well as other aspects of vaccine development.

In the selection of subjects, in the recruitment of families efforts should be made to assure that the risks as well as the benefits are widely distributed and there is public scrutiny of the whole selection process. The use of prisoners as test subjects violates the requirement for public observation and openness in research as well as using persons with limited autonomy. Certainly discrimination on a basis of sex, race, geography, nationality or economic status is unfair and investigators should be forced to offer opportunities to participate to many groups. The inducement to participate in the early development of biologicals by remuneration of the subjects is also controversial.

On the one hand society has in general accepted the remuneration of those who play a role in national defense, even though their risks are higher than the rest of society. Certain occupations with much risks have always been condoned or even exalted such as that of the test pilot and are well remunerated. Paying subjects to assist in developing biologicals to defend the public against disease does not seem inappropriate when viewed from that perspective. On the other hand those in need of money are far more likely to volunteer if fairly well paid and so the risks would fall on the lower socioeconomic group with the benefits being gained by the upper.

If biologicals can be considered public resources then a draft of subjects might be more equitable than a mercenary "health force."

The compensation of families for the participation of their children already encourages the exploitation of the less powerful in the family and cannot be condoned.

If in the development of vaccines and if in their use in field trials adverse reactions occur which produce harm should compensation for injury be provided? Again the principle of justice seems applicable.

If biologicals are a public resource, as stated, then the costs of development should be borne by the public. The cost of repairing or caring for injury and compensation for the discomfort and disability of the subject should be a part of the development cost.

On the other hand, if an adult fully comprehends the risks and knowingly takes those risks out of a sense of obligation to others, is compensation for injury owed to that individual or should it be given to recognize his altruism?

Regardless of the answer, whether or not compensation is given must be a part of the information that is provided during the consent procedure and the subject should know unequivocally what is intended as regards compensation. One of the serious deficiencies in the swine flu information piece that informed the vaccine recipient was that relating to compensation.

The assumption of responsibility of the U.S. government in lieu of that by manufacturers was interpreted by many as a commitment to compensate for adverse reactions to the vaccine. The many law suits now pending are a consequence.

In summary ethical considerations underlie many aspects of the development of biologicals and the use of cells. Clear cut unequivocal answers to many of the dilemmas cannot be given. Only imperfect approximations are possible. However, since the importance of disease prevention is paramount to prevent harm it is critical that the fullest participation of the general public not just its scientist representatives be carried out. By openness, by a full presentation of the facts and the arguments reasonable decisions favoring no one segment of society should emerge.

The track records of the National Commission for the Protection of Human Subjects and of the many institutional review boards although far from perfect are offered as evidence. The progress of Science, contrary to the fears of many is not seriously impeded and the rights, particularly of the powerless are not seriously abused.

REFERENCES

1. Appendix B-2 Report and Recommendations of the National Immunization Work Groups submitted to the Office of the Assistant Secretary for Health. JRB Associates, Inc. March 15, 1977.
2. National Immunization Conference, Policy Perspective, HEW., P.H. S. November 1976. DHEW Publication # (OS) 77-50039 p.8
3. Report and Recommendations: Research Involving Children. The National Commission for the Protection of Human Subjects of Biomedical and Behavioral Research, DHEW Publication No. (OS) 77-0004.

REPORT OF AD HOC COMMITTEE ON KARYOLOGICAL CONTROLS OF HUMAN CELL SUBSTRATES

Lake Placid, New York

October 26, 1978

INTRODUCTION

A conference on the Standardization of Cell Substrates for the Production of Virus Vaccines was jointly sponsored by the World Health Organization and by the International Association of Biological Standardization at Geneva in 1976. Participants at the meeting agreed that recommendations then current regarding karyological control of human cell substrates should be reviewed and revised. Extensive experience using human diploid* cells in vaccine manufacture as documented at that symposium indicated that certain karyological requirements of existing standards should be eliminated or modified. An Ad Hoc Committee appointed by the Chairman of the conference, Dr. Frank Perkins, was charged with this task; however, since opinions varied widely regarding modification this committee requested the organization of a karyology workshop.

*Diploid (2n) is defined as the exact double of the haploid (n) chromosome complement. As becomes evident from the subsequent discussion, a small proportion of cells in a diploid cell line have other chromosome numbers, mainly tetraploid (4n). These deviations from strict diploidy are also encountered in cytological materials prepared from tissues *in vivo*. The term euploid means all, or any exact multiple, of the haploid complement; therefore, the term diploid is considered better suited to describe the chromosomal content of the cells of these lines (1,4,8).

STATEMENT ON KARYOLOGICAL MONITORING

by

IABS/WHO Ad Hoc Committee on Karyology
Geneva, 13-15 December, 1976

After brief deliberation, this committee concludes that no effective concensus as to specifics could be reached in the time available concerning modifications of the present requirements for karyological monitoring of cell substrates for viral vaccine production. It is recognized that existing requirements for production lot monitoring probably should be modified, in the light of the new extensive experience with approved diploid human cells. It is therefore recommended that the means be sought for organizing a workshop conference of those concerned with karyological aspects of safety requirements for viral vaccine production. This workshop should be held within three to six months.

As a preliminary to this proposed workshop, the members of this Ad Hoc Committee have arranged to exchange individual assessments of these problems. The Bethesda Conference in September, 1976, the present conference, and the working proposal submitted by Dr. P. Jacobs now provide a basis for a consideration of details of proposals for modification of the requirements for production lot monitoring. This interchange should provide a foundation for final agreement which we feel can only be attained under the conditions of a one or two day workshop conference.

The objectives of this workshop should also include recommendations of principles involving both characterization and monitoring requirements in applications to unique problems which are anticipated, such as the projected use of cells from non-human species or human cells with growth or karyological characteristics which may differ from diploid human fibroblasts.

Specific suggestions, information and advice related to these goals are solicited from the participants of this Joint WHO/IABS symposium and from other interested parties.

Since direct support for organization of such a workshop was not available, review and revision of existing standards was delayed until October of 1978 when it was arranged that the Ad Hoc Committee members should confer in a session attached to the Second Annual W. Alton Jones Cell Science Center Symposium: Cell Substrates and Their Use in the Production of Vaccines and Other Biologicals. This meeting was held at Lake Placid on October 26, 1978. The committee members and observers are listed in the appendix to this report.

The following revised karyology control recommendations are offered with the hope of universal adoption by control authorities in order to promote uniform official requirements throughout the world.

CHARACTERIZATION OF NEW DIPLOID CELL LINES

It is recommended that at least four samples be examined at approximately equal intervals over the lifespan of the cell line during serial cultivation. Each sample should consist of 1000 metaphase cells.

It is also recommended that photographic reconstruction be employed in the preparation of chromosome banded karyotypes of 50 metaphase cells per 1000 cell sample using either G-banding or Q-banding techniques. This constitutes five percent of the sample and no specific limits for acceptability are yet recommended. The incidence of karyotypic abnormalities (pseudo-diploidy, inversions, translocations, etc.) that are detectable with the greater resolution provided by banding should be evaluated upon the accumulation of a larger data base than is presently available.

MANUFACTURER'S WORKING CELL BANK MONITORING

The Manufacturer's Working Cell Bank (MWCB) consists of all of the cellular material derived by serial subculture from an approved diploid cell line seed bank. In normal practice such a seed culture (or ampoule) is issued to manufacturers at or near the eighth population doubling level (PDL). This is expanded by serial subculture until a PDL selected by the manufacturer at which point the cells are combined into one or more pools and preserved cryogenically to form the MWCB. Karyological analysis is to be carried out on cultures representative of <u>each pool</u>.

It is recommended that the location and identity of individual ampoules comprising the MWCB shall be thoroughly documented. All other cells shall be excluded from the physical facilities used for cryogenic preservation of the MWCB. The maximum interval of time between karyological monitoring of a particular pool and eventual production use of any portion of that pool shall be left to the discretion of the control authority.

Metaphase preparations shall be made from a sample of cells representative of each pool of the MWCB. At least 500 metaphase cells shall be examined and observations should conform to the limits as recommended for each karyological parameter ascertainable with conventionally stained preparations (Table 1). This is to be done at a PDL equivalent to the final subcultivation level at which vaccine or other biological product is prepared, or at any PDL thereafter. This requirement provides cytogenetic data from various laboratories for comparison and a means for evaluating the cytogenetic competence of personnel, the genetic stability of the cell line, and adequate manufacturing practices.

PRODUCTION MONITORING

It is recommended that karyological sampling of the production lot, at the level of vaccine production or cell product extraction, be reduced to the preparation of a single representative karyotype by either conventional staining or by G- (or Q-) banding technique. This single karyotype is required in order to confirm the identity of the approved cell line actually used, in terms of its species, sex, and any known polymorphic feature. Any alternate technique of equivalent sensitivity and reliability for cell identification, such as HLA determinations or isozymes, etc., may be substituted for this requirement, at the discretion of the control authority.

Metaphase Preparations

Permanent stained slide preparations for examination of metaphase chromosomes of diploid cells may be made according to any of various widely published procedures for conventionally stained preparations, for G- or Q-banded preparations, and for C-banded preparations.

Conventionally stained preparations are required for the assessment of: frequency of chromatid and chromosome breaks, structural abnormalities, hyper- and hypodiploidy, polyploidy, and "other" abnormalities such as attenuated primary and secondary constrictions. Either G-banding or Q-banding techniques are required for karyotyping, and C-banding is recommended for a full definition of any C-polymorphism present which may serve as an identifying marker of a candidate cell line. Conventions of terminology for human chromosomes are provided in the Paris Conference (1971) proceedings (6). With all techniques the overcondensation of metaphase cells must be avoided; exposure to colchicine or to commercially available analogues should not exceed 1-^{10} M final concentration, nor exceed fours in time. The actual procedures adopted must be approved by the licensing authority in the course of demonstrating good laboratory practices and competence by the manufacturer's personnel.

TABLE 1

Recommended Minimal Requirements for Kayological Parameters for Human Diploid Cell Lines, Revised 1979

ABNORMALITY	Expected Frequency (Mean values, WI-38)***	UPPER LIMITS OF ACCEPTABILITY* 1000-Cell Sample	500-Cell Sample
Chromatid & Chromosome Breaks	3.51%	47/1000	26/500
Structural Abnormalities	∿1.%	17/1000	10/500
Hyperdiploidy	0.37%	8/1000	5/500
Hypodiploidy	n.a.	180/1000	90/500
Polyploidy	2.7%	30/1000	17/500
Other Abnormalities**	n.d.	---	---

* Upper fiducial limit @ 95% confidence (Poisson).

** At discretion of control authority

*** Normal values as established from conventional metaphase data from WI-38 cells at 25-35 PDL (majority @ 27-33 PDL), Chatham Conference, 1971. Values for breaks, structural abnormalities and hyperdiploidy based on 500 samples of 100 cells each. Value for polyploidy based on 46 samples of 300 cells each. All values tested for homogeneity and only that for "structural abnormalities" was not homogenous (range 0.0-1.3%). Provisional value of ∿1% accepted as approximate mean frequency.

n.a. Not applicable

n.d. Not determined

Karyological Parameters

a) Chromatid and Chromosome Breaks. During the examination under oil immersion optics of each cell of a sample, all unequivocal discontinuities of the chromatids or chromosomes are to be tabulated on the basis of "cells with 1 or more breaks."

It is recognized that the allocation of lesions into distinct categories can be ambiguous in the borderline cases. To reduce subjectivity in scoring, it is recommended that any discontinuity of the chromatid(s) that is equal to or greater than the width of a chromatid be considered unequivocal and recorded as a "break." Lesser lesions may be tabulated as achromatic gaps, but for accessory information only. Thus, "breaks" (chromatid or chromosome) as here defined would include aligned as well as non-aligned discontinuities and would also include acentric fragments, whether the origins were evident or not.

b) Structural Abnormalities. Within this category are included those abnormal configurations considered typical such as muticentrics, dicentrics and rings, (whether acentric or centric) as well as chromatid exchanges and stable "marker" chromosomes. In this context markers are included as possible indicators of prior exchange events. Very small chromosomes ("minute") which possess a centromere are considered to be markers and are also scored in this category. The term "minute" is used also for bipartite chromatin bodies which may represent terminal acentric fragments, interstitial deletions, or the products of other events (double minutes or DM's). This type of "minute" is to be scored with Chromatid and Chromosome Breaks (see (a)). It is emphasized that the description or identification of the type of aberration should be as specific as possible to permit a more valid comparison of data.

c) Hyper- and Hypodiploidy. An exact count of the chromosomes in each metaphase cell of the sample should be obtained. Metaphases may be selected for general cytological suitability at low power magnification (ca. 100 X), but once selected a cell may not be excluded from compiled data even if an exact count is not feasible. All aberrations observed must be entered into appropriate categories. This procedure minimizes possible bias in the selection of cells for technical quality. The counting of chromosomes and observations for the presence of aberrations must be performed under "oil immersion" optics (ca. 1000 X). Since cells observed with less than the diploid number of chromosomes (hypodiploidy) predominantly result from the technical artifact of overspreading, it is recommended that no preparation be used in which the apparently aneuploid counts exceed and arbitrary limit (Table 1). Different cytological preparations from the same sample of cells may be substituted for any preparation which fails to meet this requirement for hypodiploidy. The limit acceptable for hyperdiploidy is based upon the mean of pooled data from the WI-38 cell line.

It is recommended that analyses of conventionally stained karyotypes, including photographic reconstruction, shall be performed where possible on all cells with aneuploid counts. The chromosomes or groups having "missing" or "extra" members should be determined.

d) Polyploidy. Appropriate randomized observations on a minimum of 500 unselected metaphase cells representing each sample should be made using medium power optics (ca. 250 X) for the determination of polyploidy (tetraploidy) frequency in the dividing cell population. Only the general ploidy level for each metaphase need be determined, i.e. each cell can be classified reliably as 2N, 4N, etc. by an approximate count, or by using higher power optics if any doubt remains.

e) Other abnormalities. Whole chromosomes or regions of metaphase chromosomes may appear despiralized and are to be considered abnormalities, except where this is the expression of one of the known secondary constrictions near the centromere of chromosome pairs Nos. 1, 9 and 16. Extreme attenuations of the primary or secondary constrictions should be scored as abnormalities. Both of these types of aberrant chromosomes are well described (9) and are to be scored in this category.

Karyotype Analysis

In the characterization of new diploid cell lines, G- or Q-banded chromosome karyotypes should be prepared from a number of metaphase cells which equals five percent of the sample required for the various parameters determined from conventionally stained material. It is recommended that reference coordinates for relocating each cell karyotyped be recorded to permit direct microscopic evaluation during karyotype preparation from photographic prints or for review in consultation with licensing authorities. Chromosomal variants may be present in all cells of some normal individuals and are considered to be polymorphisms. Examples include: variations in size of the constitutive heterochromatin in the long arms of chromosome Nos. 1, 9 and 16; size variation of the long arm of the Y chromosome, and enlarged satellites. At the present time these consistent and heritable minor variant chromosomes are presumed not to indicate pathology but should be thoroughly described as part of the characterization of any cell line to be considered for use.

LIMITS OF ACCEPTABILITY

Proposed limits for acceptability for the appropriate size samples are presented in Table 1. The determination of the expected mean frequency or background level of abnormalities, is based upon cumulative data obtained from presumably normal human diploid cells, WI-38 (7,5). In practice, if the stated limit were exceeded for a particular karyological parameter in one sample, the manufacturer retains the option of extending the size of the total number of cells examinated in increments which are not less than the original sample size (500 or 1000 cells). The results which must then conform to statistical expectation for 95% confidence in evaluating the acceptability of such an enlarged total sample.

GENERAL CONSIDERATIONS

Although our experience in karyological control of human cell substrates was derived from use of the fibroblast cell line WI-38, these revised requirements do not specifically exclude any cell line on the basis of cell type or pattern of lifespan. It is recognized that human cell lines other than fibroblast-like lines may be acceptable to control authorities even though their karyological characteristics can differ from those of human diploid fibroblast cell lines. The limits of acceptability recommended in this report should be viewed as explicitly applicable to all human diploid fibroblast cell lines* and as possibly useful in determining the acceptability of other human cell lines. Similarly, diploid cell lines derived from non-human species may be acceptable to control authorities even though their karyological characteristics differ from those of human diploid fibroblasts. The characterization studies of such non-human cell lines should follow the general guidelines as presented for human cell lines.

There was general agreement that for the purposes of karyological monitoring manufacturers should not necessarily be restricted to a specific PDL. This is considered a matter to be determined by consultation between manufacturer and control authority, with the burden on the manufacturer to demonstrate the basis for the PDL limit at which production should be allowed. In principle any such limitation should be based upon the stage at which a given diploid cell line might fail to meet karyology standards. Other criteria for acceptance of a diploid cell line, such as tumorigenicity tests, may preclude the use of the cells at a given stage in their life history even though the karyology is acceptable.

The relevance to product safety of karyologic monitoring of cell substrates in the production of non-replicating biological products such as interferon is open to question. The potential risks of using karyologically abnormal cells are related to the possible con-

---

*At present, in addition to WI-38 only the human diploid fibroblast cell line MRC-5 meets these requirements for karyolgoical control (2,3).

tamination of the product with cellular nucleic acid. Control authorities should therefore consider the purification procedures used in the manufacturing process and the confidence with which one can exclude the presence of cellular nucleic acid in the final product. Current technology may allow the elimination of cellular nucleic acids from non-replicating biological products produced in cell culture systems, in which case karyologic monitoring of the cell substrate is considered unnecessary.

REFERENCES

1. Federoff, S. 1967. The proposed usage of animal tissue culture terms. J. Nat. Cancer. Inst. 38:607-611.
2. Jacobs, J.P., Jones, C.M. and Baille, J.P. 1970. Nature 227: 168-170.
3. Jacobs, J.P. 1977. Joint WHO/IABS Symposium on the standardization of cell substrates for the production of virus vaccines, Geneva, Dec. 1976. Develop. Biol. Standard., 37:155-156. Basel: S. Karger.
4. Levan, A. and Muntzing, A. 1963. Terminology of chromosome numbers. Port. Acta Biol. 7:1-16.
5. Moorhead, P.S., Nichols, W.W., Perkins, F.T., Hayflick, L. 1974. J. Biol. Standard. 2:95-101.
6. Paris Conference 1971. Standardization of Human Cytogenetics. Birth Defects: Original Article Series (Hamerton, Jacobs and Klinger, eds.), VIII, No. 7, p.46. New York: The National Foundation.
7. Perkins, F.T. 1971. Minutes of the Eighth Meeting of the Committee on Cell Cultures. International Association of Biological Standardization Cell Culture Committee.
8. Rieger, R., Michaelis, A. and Green, M.M. 1968. A Glossary of Genetics and Cytogenetics, 3rd edition, p. 507. New York: Springer-Verlag.
9. zur Hausen, H. 1967. J. Nat. Cancer Inst. 38:683-696.

1976 WHO/IABS Ad Hoc Karyology Committee

Dr. P. Moorhead, Chairman
Children's Hospital of Philadelphia
34th and Civic Center Boulevard
Philadelphia, PA 19103

Dr. E. M. Earley
Bureau of Biologics, FDA
8800 Rockville Pike
Bethesda, MD 20205

Mr. K. Frieling
Paul-Ehrlich-Institut
Paul-Ehrlich-Strabe 44
D-6000 Frankfurt/M.70
West Germany

Mr. P. Jacobs
National Institute for Biological Standards and Control
Holly Hill, Hampstead
London NW3 6RB England

Dr. J. Litwin
State Bacteriology Laboratory
S-105 21 Stockholm
Sweden

Dr. R. Netter
Laboratoire National de la Sante
25, Boulevard Saint Jacques
F-75014 Paris France

Dr. J. Petricciani
Bureau of Biologics, FDA
8800 Rockville Pike
Bethesda, MD 20205

Dr. A. von Seefried
Connaught Laboratories Ltd
1755 Steeles Avenue West
Willowdale, Ontario M2N 5T8
Canada

Observers for 1978 Lake Placid Session of Ad Hoc Karyology Committee

Dr. O. Andzhaparidze
Moscow Research Institute for
Viral Preparations
1 Dubrowskya 15
Moscow, USSR

Dr. M. Beck
Institute of Immunology
Rockefellerova 2
Zagreb, Yugoslavia

Dr. S. Dzagurov
Tarassevich Institute
USSR Ministry of Health
Sivtsev-Vrazhek 41
Moscow G-2 USSR

Mr. R. Ferris
Wellcome Research Laboratories
Langley Coutrt
Beckenham, Kent BR3 3BS UK

Dr. H. Mcamorine
Connaught Laboratories Ltd
1755 Steeles Avenue West
Willowdale, Ontario M2N 5T8
Canada

Dr. R. Mauler
Behringwerke AG
D-355 Marburg/Lahn
West Germany

Dr. W. A. Nelson-Rees
University of California
Naval Biosciences Lab
Oakland, CA 94625

Dr. S. Pathak
Section on Cell Biology
M. D. Anderson Hospital
Houston, Texas 77030

Dr. K. R. Schell
Swiss Serum & Vaccine Institute
P O Box 2707
3001 Berne, Switzerland

Dr. W. J. Thomas
Salk Institute
Government Services Division
Swiftwater, PA 18170

Dr. V. Turner
Institute of Immunology
Rockefellerova 2
4100 Zagreb, Yugoslavia

Dr. A. Whitaker
Wellcome Research Laboratories
Langley Court
Beckenham, Kent BR3 3BS UK

Dr. D. P. Yang
Wyeth Laboratories, Inc.
P O Box 8299
Philadelphia, PA 19101

INDEX

MIX
Papier aus verantwortungsvollen Quellen
Paper from responsible sources
FSC® C105338

If you have any concerns about our products,
you can contact us on
ProductSafety@springernature.com

In case Publisher is established outside the EU,
the EU authorized representative is:
Springer Nature Customer Service Center GmbH
Europaplatz 3, 69115 Heidelberg, Germany

Printed by Libri Plureos GmbH
in Hamburg, Germany